Transforming the Task with Number Choice

Kindergarten through Grade 3

Tonia J. Land
Corey Drake
Molly Sweeney
Natalie Franke
Jennifer M. Johnson

NATIONAL COUNCIL OF
TEACHERS OF MATHEMATICS

Library of Congress Cataloging-in-Publication Data

Transforming the task with number choice / Tonia J. Land [and four others].
      pages cm
Includes bibliographical references.
ISBN 978-0-87353-742-1
1. Number concept. 2. Mathematical recreations. 3. Mathematics—Study
and teaching. I. Land, Tonia J.
QA141.15.T73 2014
372.7—dc23
                                                          2014025495

The National Council of Teachers of Mathematics is the public voice of mathematics education,
supporting teachers to ensure equitable mathematics learning of the highest quality for all students
through vision, leadership, professional development, and research.

Printed in the United States of America

CONTENTS

Foreword

I love the goal of this book! Number selection is foundational to effective mathematics instruction, but surprisingly few resources exist for learning to choose numbers that are responsive to individual children's needs. Number selection is also hard! I know that I have had the experience of posing a problem only to realize—too late—that the numbers I chose were not working as intended, but instead, were leaving children confused, underchallenged, or overchallenged.

In *Transforming the Task with Number Choice*, we are taken on a journey into this important and difficult topic at the elementary school level. Extensive and rich examples of mathematical problems and classroom scenarios offer a glimpse into how teachers can thoughtfully analyze and plan for a range of problem types according to their mathematical features and the ways children are likely to interact with them. We learn about characteristics of productive number choices and how to structure lessons to capitalize on these characteristics. We also learn that to select productive number choices that move individual children's thinking forward, we first need to attend to the details in their strategies and the understandings reflected in those details.

This book is timely! Explicit links are drawn to the Common Core State Standards in Mathematics—including both the Standards for Mathematical Practice and the Standards for Mathematical Content. The authors help us envision how to implement these standards by providing sample problems with strategically selected numbers as well as sequenced sets of numbers to address particular standards. In addition, we learn how we can use a single problem, often presented with multiple number choices, to address several standards at the same time.

Throughout the book, the authors draw on research on children's mathematical thinking in ways that complement many respected projects and resources focused on the importance of children's reasoning, how that reasoning develops, and how instruction can build on that reasoning. I particularly appreciated the authors' notions of "just right goals" and "just right numbers" for individual children. Although most would agree that lesson goals are critical, the authors recognize that the situation is more complex, and a single goal for all children is generally insufficient. Instead, they suggest that teachers work to select "just right goals" and "just right numbers" for individual children—goals and numbers that challenge but also enable each child to access problems and feel successful. Children can learn to choose these "just right numbers" as well.

Tailoring instruction to the needs of individual children is challenging, especially because instruction occurs in a classroom with many children. However,

the authors provide us with tools to engage in this type of responsive instruction. Through many examples and practical suggestions for using strategic number selection to adapt or fine-tune lesson goals and curriculum, we gain a vision for how we can simultaneously address the needs of individual children. This type of teaching is incredibly complex, yet I strongly agree with the authors that teaching based on responsiveness to individual children is critical and an issue of equity—one that allows us to move beyond the "low, medium, and high" approach for differentiating instruction to one that instead recognizes and builds on the strengths of each child.

Traditionally, this important process of number selection has been hidden in practice and even in many teacher resources, but the authors convince us that we can greatly benefit from making it visible. Through their detailed descriptions of teachers' analyses and reasoning about number selection, we not only learn from the examples but also know how to begin to engage in conversations with colleagues around these ideas. Only through continued dialogue about number selection will we ensure productive number choices for each child year after year. *Transforming the Task with Number Choice* will be an invaluable resource in these discussions. We can find sample problems, numbers, and sequenced sets of numbers to accomplish a variety of mathematical goals as well as questions to ask ourselves when we try to reason through the demanding work of instructional decision making. Thank you to the authors for starting the conversation and providing such a timely, accessible, and useful resource!

Victoria R. Jacobs
University of North Carolina at Greensboro

Preface

The five of us—Tonia Land, Corey Drake, Molly Sweeney, Natalie Franke, and Jennifer Johnson—came together through a research project funded by the National Science Foundation. The purpose of the original research was to discover and understand how teachers use curricular resources in instruction. Through this work, while it was not the stated focus of our study, Corey and Tonia developed a particular interest in the many different ways in which Jenny, Molly, and Natalie used number choice in their classrooms. They recognized the benefits of number choice and used it in ways that made the mathematics accessible, equitable, and differentiated.

Through the original research, we became aware of the lack of number choices in curriculum materials, and that made us recognize that not all students are able to access the mathematics of a problem at their level. We began thinking about and discussing the following questions: What types of numbers allow an entry point for all of the students? Does the sequence of numbers in the progression influence how a child solves the problem? Does size or the relationship between the numbers help develop relational thinking? Do the numbers allow the students to vary their strategies and utilize the properties of operations? With these questions in mind, we began our journey of crafting purposeful number choices and progressions to meet the needs of students and improve teaching practices.

Through mutual support and cooperation, the five of us began taking a deeper look into the Common Core Standards, examining our classroom practices, and investigating student work.

We share a common belief about mathematics teaching and learning in that first and foremost students should be doing the mathematical work in a classroom. Children best engage with this mathematical work when presented with rich, worthwhile mathematical tasks guided by teacher questioning.

The Teaching Principle in *Principles and Standards for School Mathematics* (NCTM 2000) states the following about "worthwhile tasks":

> In effective teaching, worthwhile mathematical tasks are used to introduce important mathematical ideas and to engage and challenge students intellectually. Well-chosen tasks can pique students' curiosity and draw them into mathematics. ... Regardless of the context, worthwhile tasks should be intriguing, with a level of challenge that invites speculation and hard work. Such tasks often can be approached in more than one way ... which makes the tasks accessible to students with varied prior knowledge and experience. (NCTM 2000, pp. 18–19)

In this book, we suggest that the work of teaching elementary mathematics with and through worthwhile tasks is more manageable—and more accessible to all students—through a focus on number choice. We define *number choice* as the strategic use of numbers and number combinations in the context of problem-solving tasks. This book provides examples of and strategies for using number choice, based on our teaching experiences and research, to "engage and challenge students" with problems that "can be approached in more than one way."

Multiple resources, including textbooks, curriculum materials, supplemental materials, and professional development books for teachers, exist to aid in the selection, adaptation, and enactment of worthwhile tasks. However, few of these resources provide specific information about how to choose or adapt numbers within these problems. Even fewer suggest strategies for using productive numbers in the context of worthwhile tasks in ways that are responsive to children's mathematical strengths and needs. To be clear, it is often the case that the number choices in these resources are very thoughtful and strategic, and are designed to address specific mathematical goals and to progress logically over time. However, the rationales for these number choices are often not shared with readers, making it difficult for teachers to apply the rationales to other problems. (An exception is *Number Talks* (Parrish 2010) that provides many examples of strategic number choices, as well as some insights into the rationales behind those choices, specific to the number talk instructional structure.) Furthermore, single number choices are not likely to be responsive to all twenty-five or thirty children in any given classroom, and therefore, there is a need for strategies for choosing multiple numbers for single problems. This book is intended to address these gaps by providing explicit methods and examples for using number choice to address the mathematical content, strategies, and practices in the Common Core State Standards for Mathematics.

This strategic approach to number choice is closely connected to our more general view of successful mathematics teaching and learning: Children make sense of mathematics through engaging with well-written, or "worthwhile," tasks and problems, and this sense-making process then supports the development of computational fluency. Our approach is supported by our own experiences as teachers and researchers as well as by prior research (e.g., Carpenter et al. 1999; Fosnot and Dolk 2001a, 2001b, 2002). We feel that successful mathematics teaching and learning can be accomplished through the use and adaptation of well-designed instructional resources, including those published by Carpenter and colleagues and Fosnot and colleagues as well as recent Common Core Standards–based curriculum materials. However, we suggest that these resources must be adapted to be responsive to individual children and groups of children, and that a focus on number choice is an effective strategy for achieving this kind of responsiveness. In this book, we provide examples and strategies to show how a focus on number choice can not only address multiple aspects of the Common Core Standards (or others) but also meet the needs and strengths of individual students.

The Common Core State Standards (NGA Center and CCSSO 2010) provides frameworks for organizing addition/subtraction and multiplication/division problems as part of the standards for mathematics content. The frameworks describe and provide examples for all the common addition/subtraction and multiplication/division situations. For example, equal group problems can have either the product unknown, group-size unknown, or number of groups unknown. These frameworks provide significant direction for selecting, adapting, and writing problems of different types, in much the same way that the research on Cognitively Guided Instruction (Carpenter et al. 1999) has done. However, the standards provide minimal direction for selecting numbers for problems. In first grade, students are to add and subtract within 20. Students in second grade are to add and subtract within 100. In third grade, students are to multiply and divide within 100. These recommendations lack specificity in terms of—

- the progression of number choices within these broad landmarks (e.g., 20 and 100);
- the properties of numbers that might be considered in choosing or adapting numbers for problems (e.g., decade or non-decade numbers that combine to make ten and doubles); and
- the ways to introduce and sequence numbers to build students' mathematical understandings.

In addition to mathematics content, the Common Core also calls for students to develop specific mathematical strategies (e.g., strategies based on place value) and mathematical practices (e.g., reason abstractly and quantitatively). However, again, the standards do not provide guidance in how to choose or adapt numbers that would support students' development of mathematical strategies or practices. We have written this book to provide that guidance, and we illustrate how a focus on number choices can be productive in addressing three primary aspects of the Common Core: 1) mathematics content, 2) mathematics solution strategies, and 3) mathematics practices.

Because of our own work on number choice and the many questions we have received from teachers about number choice, we realized there were few resources to support teachers in incorporating purposeful number choice into their mathematics teaching. And so the five of us collaborated for more than a year to create this resource to help others on their number choice journey.

The book is divided into seven chapters. Chapter 1 lays the groundwork for understanding and using number choice by analyzing the number choices of a particular problem and rewriting them to address the development of mathematics content, strategies, and practices as described in the Common Core. Chapter 2 explains the multiple number choice structure and how it can be used to differentiate instruction. Chapter 3 discusses how to use number choice and curriculum materials to support the mathematical practices. In chapters 4 and 5, we return to previous problems and examine additional examples, concentrating the discussion on mathematical strategies. Chapter 6 builds on the work of Carpenter and colleagues (2003) and Jacobs and colleagues (2007) in relational thinking by discussing how number choices in contextualized word problems can develop relational thinking skills. In chapter 7, we provide various tools for assessing students and their engagement with number choices. Throughout each of the chapters, we discuss how the work with number choices can be used with curriculum materials.

Ultimately, our goal is for students to become fluent in mathematics and to be flexible thinkers who choose a strategy based on the numbers in a problem, seeking and using relationships in numbers when problem solving. Fosnot and Dolk describe mathematicians and people with good number sense as ones who "would look at the numbers first to decide on a strategy" (2001, p. 115). When teachers pose purposeful number choices, they provide students with opportunities to become confident and successful mathematicians. We hope this book will be helpful as you begin introducing number choices to your students. Enjoy!

Acknowledgments

This work was supported, in part, by the National Science Foundation under Grant Nos. 0643497 and 1158860 (Corey Drake, PI). Any opinions, findings, conclusions, or recommendations expressed in this material are those of the authors and do not necessarily reflect the views of the National Science Foundation.

We would like to thank the administration and our colleagues at Jefferson Elementary and the Downtown School in Des Moines Public Schools.

We would also like to thank our husbands for taking care of the home front on many Sunday nights, and to our kids for their support and understanding as we spent time away from them.

Chapter 1

Focus on Number Choice

In our work, we have found that careful selection of number choices can be a powerful instrument for developing students' understanding and for meeting multiple aspects of successful mathematics teaching. Consider the Pennies problem written for a third-grade classroom from *Investigations in Number, Data, and Space* (TERC 2012, p. 71):

- Last month, Luis collected 97 pennies. This month, he collected 143 pennies. How much money does Luis have now?

What do you notice about these number choices?

We noticed a few things:

- There is a two-digit number plus a three-digit number.

- The two-digit number comes first.

- One of the numbers is close to 100.

- The answer (sum) will go over another hundred.

- The 7 (from 97) and 3 (from 143) can be combined to make a ten.

- The 7 from 97 could be added to 143 to make 150; or the 3 from 143 could be added to 97 to make 100.

You might have noted some other characteristics of this number choice. In general, we try to look for the following characteristics of number choices, including those in both the problem and the solution:

- Size of the numbers

- Complexity of the numbers

- Proximity of numbers to "landmark" numbers (e.g., decade numbers, 100, 25; TERC 2012)

- Ways in which the numbers can be combined (e.g., 7 from 97 and 3 from 103 can be combined to make a ten)

- Relationships among the numbers (Are the numbers close together or far apart? Do they lend themselves to decomposing and recomposing in particular ways?)

These various aspects of the numbers in a problem help us identify the mathematical goals that might be addressed as students work to solve the problem.

Directing our attention on number choice also helps teachers think about how students might engage with a problem. Some students might struggle with the numbers in the Pennies problem above, while others may quickly arrive at a solution—illustrating the need for alternate or additional number choices. The number choices above are interesting in that they can be manipulated in various ways to arrive at a solution. For instance, as already noted, the 7 and 3 can be combined to make a ten, or the 97 could be changed to 100. But will students notice those aspects of the numbers? In order for students to use and understand these strategies based on place value, they might first need to experience number choices that promote making a ten alone, for example, (8, 2), and those that are part of larger numbers, for instance, (14, 46). They might also need experiences for going over a hundred and for using the commutative property to solve problems.

To illustrate how a focus on productive number choice might work, we rewrote the numbers in the Pennies problem to support the development of mathematics content, strategies, and practices as described in the Common Core while meeting the needs and strengths of individual students:

- Last month, Luis collected _____ pennies. This month, he collected _____ pennies. How much money does Luis have now (TERC, 2012, p. 35)?

  (57, 43)   (97, 43)   (97, 143)   (107, 143)   (290, 357)

- The first number choice, (57, 43), works for students who are not yet comfortable crossing the hundred mark. Additionally, it gives students practice making tens. For example, students solving this number choice might first make 90 (50 + 40), and then combine it with 10 (3 + 7) to make 100. Or, they might make 10 first.

- The second number choice, (97, 43), is similar to the numbers in the original problem. Again, the number choice involves potentially making a ten in the ones column, but the total in this choice, as opposed to the first choice, is greater than 100. This is also a good number choice for supporting a compensation strategy in that students could add the 3 (from 43) to the 97 to make 100, and then add 40. In addition, this number choice provides a scaffold to the third number choice in the original problem (97, 143). If students see that 143 is 100 more than 43, they can simply add 100 to their solution to 97 + 43 to solve 97 + 143.

- The third number choice, (97, 143), is from the original problem.

- The fourth number choice, (107, 143), again stresses the composition of 3 and 7 to make 10, but adds the element of seeing the relationship between 97 and 107.

- The final number choice, (290, 357), provides a challenge for students who have already learned the content and strategies supported by the original number choice. It also affords another opportunity for students to compensate. Students could change 290 to 300, add 357, and then subtract 10.

The Common Core mathematics content in this example is "adding and subtracting fluently within 1000" (3.NBT.2). Also included within that standard is the expectation that students will use "strategies and algorithms based on place value, properties of operations, and/or the relationship between addition and subtraction" (NGA Center and CCSSO 2010, p. 24). The new number choices in the Pennies problem are designed to support the use of those strategies and algorithms. By being selective in our number choices, we have provided opportunities for students to make tens, make hundreds, and use compensation. In addition, we have provided increasingly more difficult content across the number choices.

Standards for Mathematical Practice

1. Make sense of problems and persevere in solving them.
2. Reason abstractly and quantitatively.
3. Construct viable arguments and critique the reasoning of others.
4. Model with mathematics.
5. Use appropriate tools strategically.
6. Attend to precision.
7. Look for and make use of structure.
8. Look for and express regularity in repeated reasoning.

Fig. 1.1. Thoughtful number choices for worthwhile mathematical tasks can support students' engagement in the mathematical practices.

The Standards for Mathematical Practice (NGA Center and CCSSO 2010) are also promoted by the number choices above, which provide several opportunities for students to develop and engage in the majority of the practices. In fact, the use of the multiple number choice structure, in and of itself, presents repeated situations for students to engage in and develop the mathematical practices, as will be discussed in greater detail in chapter 2. As part of the first mathematical practice, the Common Core State Standards for Mathematics (CCSSM) states that mathematically proficient students "make conjectures about the form and the meaning of the solution and plan a solution pathway rather than simply jumping into a solution attempt" (NGA and CCSSO 2010, p. 6). Through the exploration and solving of a variety of number choices that are designed to have relationships across them, students are prompted to use those relationships in their solutions, and, thus, we are supporting students in the second mathematical practice. When the teacher facilitates discussions about students' solutions across the different number choices, the number choices also support students' engagement in practices 3, 4, 7, and 8. We describe the ways in which number choices support the development of the mathematical practices in greater detail throughout the book.

To further illustrate how number choices can be optimized, we provide the Fishbowl problem written by Molly for a multiage second- and third-grade classroom:

- Sam had _____ fishbowls. He had _____ fish in each bowl. How many fish did he have?

A	B	C	D
(2, 10)	(4, 20)	(3, 11)	(4, 12)
(5, 10)	(8, 20)	(6, 11)	(8, 12)

In second grade, students are to "work with equal groups of objects to gain foundations for multiplication" (2.0A). In third grade, CCSSM calls for students to "represent and solve problems involving multiplication and division" (3.0A). This problem works well for both of these standards. We have found that multiplication problems tend to be accessible at almost all grade levels and, with appropriate number choice, can build content knowledge and strategies in several areas.

For the mathematics strategy, we chose a fourth-grade standard: "Multiply a whole number of up to four digits by a one-digit whole number, and multiply two two-digit numbers, using strategies based on place value and the properties of operations" (NGA Center and CCSSO 2010, p. 29, 4.NBT.5). We chose this standard because we wanted to illustrate how work in the earlier grades can address, but also go beyond, grade-level standards. The first set of number choices (2, 10) and (5, 10) may prompt students to either draw a picture or skip-count by multiples of ten. In the last two columns, numbers were chosen to encourage students to use the distributive property in which students would break apart the tens and ones. For example, 3 × 10 = 30 plus 3 more ones would be 33.

In the latter three columns, numbers were chosen so that students would use what they know about one number choice to solve for another, for instance, to use the answer from 4 × 20 to solve for 8 × 20. Therefore, we maintain that the number choices promote the second mathematical practice. Additionally, the number choices may prompt students to "look for and make use of structure," Mathematical Practice 7, by discerning a pattern in the number choices.

Chapter 2

Multiple Number Choice Structure and Differentiation

You probably noticed that our examples in chapter 1 provided multiple number choices for each problem. We call that the *multiple number choice structure*. The structure was first introduced to Molly, Jenny, and Natalie when they participated in Cognitively Guided Instruction (CGI) professional development made available through the Iowa Department of Education. The multiple number choice structure was presented as a method to differentiate instruction, but our use of it has evolved into the ways in which we use number choice progressions to support student access to problems, to promote student engagement in the mathematical practices, and to develop students' relational thinking. In other words, one clear advantage of structuring problems with multiple number choices along a progression is the promotion of equity within the mathematics classroom. Neither students who need extra support nor those who need extra challenges are singled out for "special" instruction. By using the multiple number choice structure, the needs of all students can often be met with one task without separating students based on ability. Furthermore, this structure enables all students to have access to more rigorous or challenging mathematics and to participate together in mathematical discussions—not just those students who have been identified as needing advanced mathematics content.

We do not suggest that you begin using the multiple number choice structure with as many options as we used in the Pennies and Fishbowl problems or with sets of number choices as in the Fishbowl problem. Instead, we suggest gradually introducing number choices to students. Start by using blanks in problems where numbers would typically be supplied and by providing two number choices so students can gain experience with how the process of "plugging in" the numbers works. This first step toward the multiple number choice structure also allows you, as the teacher, to develop your expectations and procedures for recording and management of students' number choices and strategies as well as to anticipate productive responses to the student who says, "What do I do now; I'm done?"

If you use a textbook, take one problem from the text and provide an additional number choice, either up or down the progression level from the number(s) provided in the textbook problem—depending on the strengths and needs of your students. If you think the numbers in the textbook will be too challenging for many of your students, formulate a number choice that will be more accessible to those students. Similarly, if the book's number choice would be too easy, pick a more complex number. If you do not use a textbook, select a problem type from the framework provided in the Common Core Standards (Table 1, p. 88). Write the problem and choose two numbers you think would be appropriate for your students as in the example below:

- Brian had some toy cars. For his birthday, he got _____ more toy cars. Now, he has _____ toy cars. How many toy cars did Brian have to start with?

(25, 50)   (25, 110)

Once you and the students are familiar with this structure, you can include more number choices. You can also direct students to choose a particular row or column of number choices as in the Fishbowl problem. We provide additional examples of number rows and columns in chapters 4, 5, and 6.

Frequently Asked Questions

Molly, Jenny, and Natalie are expert teachers who almost always use the multiple number choice structure in their classrooms. Below, they address some of the more common questions they are asked about the structure.

How do I know which numbers are good choices? We suggest that you try different numbers and see what happens. As you work with the numbers, think about the variety of strategies your students might use. Also remember that sometimes the best learning (for teachers and students) can occur when you pick a "bad" number. For instance, Molly was working with multiplication, and she posed a problem before her students had developed an understanding of the commutative property of multiplication. Molly wanted her students to use what they knew about skip counting by fives, but placed the 5 in the "wrong" place in the following problem:

- Sam had 5 bowls of salad and he put 7 tomatoes on each salad. How many tomatoes does he need?

Needless to say, Molly's students did not skip-count by fives. It was a learning opportunity for Molly in that she realized the significance of the order of the numbers in a multiplication problem when students do not yet use the commutative property to help them solve problems. For students, the number choices were not necessarily bad; they just did not meet Molly's particular learning goal. The students were still provided an opportunity to engage in problem solving and multiplication. On the other hand, there are times when a lack of focus on number choice can cause students to really struggle with a problem. For instance, consider the following sharing problem:

- Maxine had 4 candy bars that she wanted to give to her 9 friends equally with no leftovers. How many candy bars does each person get?

This is not a good number choice when students are first exploring dividing objects among multiple people because it would be difficult for them to get the answer of 4/9. Instead, you could flip the numbers so that there are 9 candy bars being shared among 4 people. This is an easier number choice because 9 can be divided by 4 with 1 candy bar left over to share among 4 people.

How many number choices do I need? It depends on how many students are in the class and what their individual strengths and needs are, but we have found that at least four number choices are appropriate. Developing at least four number choices for each problem pushes us, as teachers, to think beyond simply providing alternatives for low-, average-, and high-achieving students and to become more purposeful with our number selections. With at least four number choices, we can target each choice to a particular strategy or way of understanding. Finally, we think it is important to make the last number choice one that extends students' thinking, makes a relational thinking connection, promotes a more efficient strategy use, or provides an experience with larger numbers. Of course, as we noted above, you might want to begin using the multiple number choice structure with only two or three choices, and then include more choices as you and your students become more familiar and comfortable with the structure.

Who picks the numbers for each student—the teacher or the student? The answer to this question really depends on the individual students in your class. We find that, for the most part, students are good at picking number choices that are appropriate for them. However, some students might need someone to say, "Today, Jasmine, I would like you to start with this number." It is important to talk with your students about choosing numbers in their zone of proximal development. It is essential that students learn to choose numbers that will challenge them, yet allow them to feel successful. We liken this process to that of teaching students to pick "just right books" (Routman 2003) by calling it "just right numbers." Students can learn to recognize that a number is too easy if they immediately have a solution strategy. The number is too hard if students do not know how to get started. It is also important to clarify to students that they do not need to solve all of the number choices, although we find that many students enjoy trying several of the choices.

Sometimes you need to push students with their number choices. For instance, Molly had one student who always completed the problem for the first two number choices but went no further. She frequently had to ask him to try a more challenging number choice. Many students like to use the first number choice as a warm-up; then they are ready to attempt the more difficult ones.

How do I ask students to share their strategies in a whole-class discussion with multiple number choices? There are many ways to organize a sharing session with a multiple number choice structure. You can pick the number choice for which the majority of the students solved the problem, and then have students share and discuss a variety of strategies related to that number choice. If a student comes up with something new or unique, or something that needs to be shared to advance the class toward your mathematical goals, you may want to talk about that strategy. You can always share for more than one number choice. In this case, it would be especially helpful to point out the relationships among number choices.

Differentiation

The multiple number choice structure, in and of itself, is a form of differentiation. Through number choice, differentiation can be provided to broad groups of students, with groups defined as students with similar strategies, understandings, or achievement levels. Number choices can also be designed for individual students. As you begin to provide multiple number choices, you will likely find that particular students will need certain numbers as evidenced by the type of strategies they use to enable them to work productively, and you will begin to develop a detailed and nuanced understanding of the kinds of numbers that work best with different mathematical strategies. (For a description of students' strategies, see Carpenter et al. 1999). In addition to the grouping of students, there are several other factors to consider when using the multiple number choice structure for differentiation, as detailed below.

Entry Points. To work productively with a problem, some students may need an entry point. Number choices can be prepared to provide students access into a problem. Many times we provide small numbers for the first number choice so that the problem can easily be direct modeled, which allows students to gain an understanding of the problem situation. After they gain understanding of the problem situation, students can then focus on using more sophisticated strategies. Another way to provide an entry point is to supply numbers with which students are familiar, or "friendly," that are not necessarily small (e.g., 50, 100).

Multiple Learning Goals. The Common Core State Standards for Mathematics (NGA Center and CCSSO 2010) provides guidelines for the mathematical content that students are expected to learn at each grade level. However, we know that not all students are ready for particular mathematical content at a given time. Therefore, students within the same classroom may need to work on separate learning goals. By providing multiple number choices, a single problem can be written with multiple learning goals in mind. Consider the following problem:

- Alice has read _____ books so far this month. Her goal is to read _____ books. How many more books does she need to reach her goal?

  (2, 10)   (5, 10)   (8, 10)   (25, 45)   (35, 75)   (55, 165)

This problem was written with two Common Core standards in mind: K.OA and 1.NBT. Standard K.OA.A.4 states, "For any number 1 to 9, find the number that makes 10 when added to the given number, e.g., by using objects or drawings, and record the answer with a drawing or equation" (p. 11). In the first three number choices, this add–join unknown problem requires students to find the number that makes 10 for 2, 5, and 8. Students who are proficient with finding number combinations that make 10 can work on another learning goal using the next three number choices. Standard 1.NBT.C.4 states that students should "Add within 100, including adding a two-digit number and a one-digit number, and adding a two-digit number and a multiple of 10 ..." (p. 16). The latter three number choices require students to find multiples of 10 in adding situations. Additionally, the last number choice goes beyond the first-grade standard in that it involves a number larger than 100.

Responding to Students. In their work, Jacobs and colleagues (2010) talked about responding to students when problem posing and thinking of students' under-standings as guidelines for problem choices. As teachers, we try to consider what we know about our individual students when choosing problems and numbers to push them into increasingly more sophisticated ways of knowing and doing math-ematics. In the next paragraphs, we describe a few ways we have responded to our students through number choices.

Consider a student who consistently direct models and does not move to a counting strategy for problems such as $15 + 6$. A productive number choice for this student might be a large number plus 1 or 2 (e.g., $83 + 1$, $84 + 2$). To direct model 83 or 84 would take quite a bit of time and, in many cases, would prompt students to look for another strategy. Adding only 1 or 2 makes the counting strategy more accessible. If the child does not move to the counting strategy, some direct ques-tions might be appropriate, such as, "What did you notice about 83 and 84?" If the student does move to the counting strategy, posing smaller numbers again, for ex-ample, $13 + 1$, might be useful to see if the student transfers the counting strategy or reverts to direct modeling.

Similarly, students might consistently use a breaking-apart-by-place strategy (e.g., $54 + 33 \rightarrow 50 + 30 = 80$; $4 + 3 = 7$; $80 + 7 = 87$) and not attempt other types of strategies based on place value (e.g., incrementing, or compensating). Conse-quently, number choices that lend themselves to those strategies may be needed. For instance, if a compensating strategy is desired, posing an addition like $99 + 56$

might prompt students to change 99 to 100, add 56, and subtract 1. In this case, it is easier for some students to compensate rather than break apart by place.

In other instances, students may need additional practice with a particular kind of strategy. For instance, if a new problem type is being introduced to students, they may need to direct model a few times. Therefore, number choices that are easier could be posed to provide opportunities for that extra practice. Likewise, if students need some support with basic fact acquisition, problems with basic fact numbers can be posed.

Number Choice Progressions

In Land and Drake (2014), the authors identify number choice progressions as "a series of numbers that increase in complexity" (p. 118). We think it is important to use number choice progressions as a method of differentiation that supports students in gaining more sophisticated ways of reasoning within a topic, such as addition. For example, consider the following problem taken from Land and Drake (2014):

- Lenore has _____ pennies and Max has _____ pennies. How many pennies do they have together?

    (30, 6)   (6, 30)   (40, 20)   (10, 68)   (45, 13)

As you can see, the number choices become increasingly more complex: a decade number plus single-digit number (30, 6); single-digit number plus decade number (6, 30); adding two decade numbers (40, 20); decade number plus non-decade number (10, 68); and adding two non-decade numbers (45, 13). The first number choice is easier than the second because students could replace the ones place in 30 with the 6. (Fig. 2.1) To do that in the second number choice (6, 30), however, would require students to know something about the commutative property. The third number choice (40, 20) is just a little more difficult than the first and second in that both numbers are two-digit decade numbers. In the fourth number choice (10, 68), both numbers are two-digits as in the third, but one is a non-decade number. To solve this number choice efficiently, students would again need to use knowledge of the commutative property. The final number choice (45, 13) consists of two-digit, non-decade numbers.

When planning to pose particular problem types to students, it can help to map out a number choice progression. Think about which numbers are easier and harder for that particular problem. Easier or harder number choices are not always determined by the value of the number. In other words, a higher number is not always a more complex number. For instance, 100 is easier to divide by 4 than 63. To

> ## Number Choice Progression
>
> • Decade number plus single-digit number (30, 6)
>
> • Single-digit number plus a decade number (6, 30)
>
> • Adding two decade numbers (40, 20)
>
> • Decade number plus non-decade number (10, 68)
>
> • Adding two non-decade numbers (45, 13)

Fig. 2.1. Number choices for the Pennies problem become increasingly more complex.

provide some guidance in developing a number choice progression, we offer several example progressions specific to various problem types below.

Equal Group–Product Unknown Problems

• Liam made _____ snowball cones. Each cone had _____ snowballs on top. How many snowballs were there altogether? (Drake et al. forthcoming)

For this problem, we contend that an appropriate number progression would start with 2, 5, or 10 snowballs per cone because 2, 5, and 10 are the easiest and most common numbers for students to count by. (Note: Students may need some time to direct model before moving on to counting strategies.) The subsequent number in the progression might either be a single-digit number of snowballs (3, 4, 6, 7, 8, or 9) or a multiple of ten. Some students will find the single-digit number easier, while others will prefer the multiple of ten. Either way, students will eventually need experiences with both. Next, posing numbers like 11 and 12 will make it easier to use their knowledge of place value to solve the problem. The number 11 is just one more than 10, so students could easily count the tens and ones separately and then add them together. Similarly, some students may find it easy to count with a 5 in the ones place. Using that same reasoning, students could then explore other multiples of ten. Employing a progression of number choices in this way would build a solid foundation for multiplication in which students could then go on to solving other groups of multidigit numbers with a single-digit number, two double-digit numbers, and two multiple-digit numbers. We provide this number choice progression, with examples, on the next page in a bulleted format:

- Groups of 2, 5, and 10: (5, 2) (8, 2) (10, 2) (15, 2); (2, 5) (5, 5) (10, 5) (15, 5); (2, 10) (4, 10) (8, 10) (10, 10)

- Groups of other single-digit numbers: (5, 3) (7, 6) (6, 9)

- Groups of some multiple of ten: (5, 20) (6, 40) (8, 70)

- Groups of 11, 12, and 15: (2, 11) (4, 11); (5, 12) (10, 12); (6, 15) (12, 15)

- Groups of a multiple of ten plus 1, 2, or 5 more: (2, 21) (4, 51); (5, 45) (10, 62)

- Groups of other multidigit numbers and a single-digit number: (2, 36) (63, 4) (125, 6)

- Groups of two double-digit numbers: (35, 12) (72, 23) (84, 65)

- Groups of two multidigit numbers: (125, 60) (856, 42)

You could pick numbers across the bulleted points above or you could focus on one type of number. If focusing on one type of number, you could also pose a problem like the following:

- There are 10 juice boxes in a carton. How many are in _____ cartons?

| 2 | 6 | 13 | 20 |

Equal Group–Group Size Unknown Problems

- If _____ friends share a bag of _____ Skittles, how many Skittles does each friend get?

In a method similar to that for equal group–product unknown problems, we would first evenly divide objects into groups of 2, 5, and 10 with no leftovers. Students likely have some knowledge of these numbers, giving them an entry point into this type of problem. Next, we recommend selecting other more difficult numbers that also involve dividing a group of objects with no leftovers. A progression of these more difficult numbers would be double-digit numbers divided by single-digit numbers that involve simpler facts (e.g., $12 \div 4$) or relate to simpler facts (e.g., $24 \div 4$, $24 \div 8$), double-digit numbers divided by single-digit numbers that are harder facts, three-digit and four-digit numbers divided by single-digit numbers, and three- and four-digit numbers divided by two- and three-digit numbers. We provide this number choice progression, with examples, below in a bulleted format:

- Dividing objects into groups of 2, 5, and 10: (2, 6) (2, 10) (2, 24) (2, 68); (5, 10) (5, 25) (5, 65) (5, 125); (10, 30) (10, 80) (10, 160)

- Double-digit numbers divided by single-digit numbers that involve simpler facts: (2, 18) (3, 18) (6, 18)

- Double-digit numbers divided by single-digit numbers that are harder facts: (7, 56) (8, 72) (9, 81)

- Three-digit and four-digit numbers divided by single-digit numbers: (4, 124) (6, 270) (8, 416) (5, 1250)

- Three- and four-digit numbers divided by two- and three-digit numbers: (20, 320) (22, 352) (325, 2600)

After posing problems with no leftovers, you can provide number choices that do have leftovers. (Note: Certain problems will not work well with leftovers, for example, the Skittles problem above because Skittles are difficult to split into fractional pieces. For our purposes here, problems with items like brownies or cookies that can be split into fractional pieces will work better.) For students who may need it, we suggest scaffolding their work with leftovers by providing numbers with no leftovers as an entry point into the problem or by mixing them in with other numbers, as in the Sharing Cookies problem below. In this problem, we would ask students to solve the numbers in a row. They could start with either the first or second row.

- Trisha and Allie are sharing _____ chocolate chip cookies. If they are shared equally, how many will each of them get?

2	4	5	8	9	12	13
30	31	50	51	66	67	83

(Land and Drake 2014)

The first number choice (2) provides an entry point into the problem, while the second number (4) provides an entry point into dividing 5 items in half. Looking at the rows, you see that almost all the odd numbers are preceded by the number before it so that students can use what they know about one number choice (4) for the next (5). When they are not needed anymore, the scaffolds can be removed.

After dividing objects into halves, students could progress to dividing objects into fourths because of the relationship between one-half and one-fourth. In a problem similar to the previous Sharing Cookies but with four people, a multiple of 4 could be posed as the entry point into the problem and other multiples of 4 could be offered followed by the next number (e.g., 12, 13, 32, 33). These numbers can be described as a multiple of 4 plus 1, or $4n + 1$. Next, multiples of 4 plus 2 or 3 ($4n + 2$, $4n + 3$) add the extra challenge of more than 1 leftover. After dividing

items among four people, because of the relationship between one-fourth and one-eighth, a possible next step could be dividing objects among eight people. Or, students could work with thirds and sixths respectively, and then other unit fractions. (For extensive work on fraction problems, see Empson and Levi 2011.) Below we provide this number choice progression, with examples, in a bulleted format:

- Dividing an odd number of objects into two groups: (15, 2) (63, 2) (155, 2) (442, 2)

- Dividing a multiple of 4 plus 1 ($4n + 1$) into four groups: (41, 4) (105, 4) (645, 4)

- Dividing a multiple of 4 plus 2 or 3 ($4n + 2$, $4n + 3$) into four groups: (42, 4) (503, 4)

- Dividing a set of objects into eight groups: (65, 8) (83, 8) (193, 8)

- Dividing a set of objects into three and six groups: (34, 3) (63, 6) (67, 3) (67, 6)

Take From–Result Unknown Problems

- _____ penguins were standing on the iceberg. _____ jumped into the water to swim. How many penguins are left on the iceberg?

    17, 8   26, 11   43, 31   76, 52   102, 59   546, 184

In this type of problem, there are several characteristics that make number choices easier or harder for students to solve: if the numbers comprise a basic fact, if the numbers involve getting to a ten or a hundred, if traditional regrouping is needed or if the numbers cross a ten or a hundred, and if the numbers are closer together or further apart (43, 31 versus 546, 184).

Try This

Choose a problem type from the framework provided in the Common Core and generate a number choice progression.

You can also choose numbers that will help students develop place-value understanding. We propose following a number progression like the one outlined below. In the examples, you will notice the features we pointed out above:

- Numbers under 10: (5, 1) (5, 4)

- Numbers more than 10 with differences less than 10: (12, 3) (12, 8) (15, 7)

- Numbers with the same digit in the ones place: (40, 10) (80, 20) (86, 26) (237, 67)

- Doubles and doubles plus/minus 1: (14, 7) (12, 6) (15, 7) (11, 5) (60, 30) (60, 31)

- Non-decade numbers minus decade numbers: (12, 10) (72, 10) (72, 40) (146, 30) (146, 60)

- Decade numbers minus non-decade numbers: (50, 18) (120, 18) (120, 58) (170, 84)

- Hundred numbers minus non-decade numbers: (100, 27) (400, 65) (400, 263)

- Non-decade numbers minus non-decade numbers: (26, 11) (26, 18) (76, 52) (83, 27)

- Three-digit numbers minus two- and three-digit numbers: (263, 99) (263, 111) (364, 232) (765, 592)

Using a Number Choice Progression

Number choice progressions can be used in different ways. You could focus on a few points of a progression or you could work on just one point of a progression and provide multiple number choices at that point. To work on several points of a progression, you could pose problems with many of the points as in the Pennies problem illustrated in figure 2.1. If students need work on a particular type of number choice, you can present multiple choices for that one type. Once students have become proficient with that type of number selection, you can stop posing it. Likewise, when the number choices you have been offering are no longer challenging, you can add other choices with more complex numbers.

Try This

Examine your textbook for a number choice progression. How specific is the progression? What properties are inherent in the number choices? How are the numbers introduced and sequenced?

If you are using a textbook, examine the number choice progression that is in the book. You might have to examine the entire chapter, unit, or series. Next, decide whether or not the progression is specific enough and provides students with adequate opportunities to engage with base-ten principles or to use multiple strategies based on place value. If there are gaps in the number choice progression, try to fill them. Then, when you are posing problems, provide students with an array of choices that fall on the progression.

Scenarios

In this chapter, we have focused on differentiation and the multiple number choice structure using number progressions. We end this chapter by providing scenarios from each of Jenny's, Molly's, and Natalie's classrooms that describe a Common Core standard they were trying to achieve as well as a strategy goal along with a problem and its number choice rationale.

Jenny's Scenario

Common Core standard (1.OA.1): "Use addition and subtraction within 20 to solve word problems involving situations of adding to, taking from, putting together, taking apart, and comparing with unknowns in all positions, e.g., by using objects, drawings, and equations with a symbol for the unknown number to represent the problem" (p. 15).

Math strategy goal: I wanted to move students from using a direct modeling strategy to using a counting on or relational thinking strategy.

Problem: Annie made _____ gingerbread cookies. She needs _____ cookies so all her friends could have one. How many more gingerbread cookies does Annie need to make?

(8, 10)   (18, 20)   (9, 12)   (9, 20)

Number choice rationale: I chose my first number set (8, 10) because I knew it would be a successful one for all the students in my room because they could all count to 10 proficiently. I also anticipated that some of my higher-level students would solve this as a known fact. I selected 18 and 20 as my next number choice because it gave students the opportunity to count on with numbers larger than 10. I also wanted to check to see if children would make a connection between solving 8 and 10 and solving 18 and 20. This promoted a discussion of "it's just 2 more" from 8 to 10 or from 18 to 20. The next number choice (9, 12) asked students to solve a problem going over a ten. Because the first number choice anchored to 10, this number choice sets students up to consider how they could anchor to 10 and then add to it. While most students counted on to 12, I looked for opportunities to talk to students about how counting on from 9 to 12 is the same as counting on 1 (to 10) and then 2 more (to 12). For the last number set (9, 20), I saw a much greater variety of strategies emerge from students. Some students reverted to counting on with cubes and tallies because "it (the difference between 9 and 20) is a bigger number." However, students who were beginning to understand how to anchor to a ten and count by tens solved this problem easily and efficiently. Discussion of the previous

number choice (9, 12) promoted this idea. A student can solve 9 and 20 as "plus 1, then 10 more; the answer is 11."

Natalie's Scenario

Common Core standard (2.NBT.1): "Understand that the three digits of a three-digit number represent amounts of hundreds, tens, and ones"(p. 19).

Math strategy goal: My goal was that students be able to recognize the number of tens in a three-digit number and use that as a means to help solve the problem.

Problem: Mrs. Franke is going to feed the turkeys. She gave 10 pieces of corn to each turkey. If there are _____ pieces of corn, how many turkeys can she feed?

40   61   109   250   517

Number choice rationale: I chose the number 40 because I wanted to give the children an accessible number to begin with. It was an entry point for all of my students. This also helped them feel successful and decreased the amount of frustration. I chose 61 because it is a non-decade number, but it's close to a decade number. I next selected 109 because it is close to 100. It provided me information about which students automatically recognized that there are ten tens in 100. The number 250 was chosen to see if students used their knowledge of the number of tens in 100 and then the number of tens in 50 to help them. This number choice also brought up discussion about dropping the zero from the ones place and realizing that the remaining digits represent the number of tens in the number (e.g., 25 tens in 250). Finally, I chose the number 517 because I wanted to see if students would recognize the relationship between the number of tens in 100 and the number of tens in 500. Also, this was an opportunity for students to see the relationship between 250 and 500.

Molly's Scenario

Common Core standard (3.NBT.2): "Fluently add and subtract within 1000 using strategies and algorithms based on place value, properties of operations, and/or the relationship between addition and subtraction" (p. 24).

Math strategy goal: Applying what they know about base ten and the relationships between numbers, students will solve the problem using mental strategies.

Problem: Mrs. Capper's class kept track of how many books they had read. In October they read _____ books. They read _____ books in November. How many books did they read altogether?

(300, 140)   (340, 220)   (420, 199)   (572, 338)

Number choice rationale: I chose the first number set (300, 140) because I wanted to see if the students could mentally add these two numbers. The second number set not only extended students into working more with hundreds, tens, and place value but also supported the use of mental math. I chose 420 and 199 because I hoped that the students would see the relationship between 199 and 200, and then move one from 420 to 199 so the problem would be 419 + 200, making it similar to the first number choice (a hundreds number plus a three-digit number). The last number choice allowed for different strategies to be used. Some students moved the 2 from 572 to add to 338 for 340; then added 570 and 340. Others noticed the 70 and 30 and added them together to create another hundred.

Chapter 3

Using Number Choice and Curriculum Materials to Support Common Core Standards for Mathematical Practice

When envisioning an elementary mathematics classroom that provides opportunities for students to achieve the Common Core Standards, teachers often wonder how to weave the *content* and *mathematical practices* into a cohesive unit. The content tells us what students should learn while the practices tell us *how* students should engage in mathematics. In "Keeping Teaching and Learning Strong," Russell (2012) contends that the mathematical practices should receive just as much attention as the content standards—and we agree. To do that, Russell suggests having "targeted, intentional, planned instruction" for each of the practices.

Teachers face particular challenges in using published mathematics curriculum materials in ways that are responsive to the mathematical thinking of the students in their classroom and that support the students in engaging in the mathematical practices. In this chapter, we share a common lesson structure for teaching mathematics using (and modifying) curriculum materials in ways that promote the mathematical practices. We present strategies for adapting curriculum materials through number choice (a specific example of adapting one *Everyday Mathematics* lesson can be found in Drake and colleagues, in press).

Lesson Structure, Curriculum Use, and Mathematical Practices

There are four basic phases to the lesson structure: number work, problem posing, student work time, and a sharing session. (See also Drake et al. forthcoming.) When planning a lesson using this structure, first select a focused learning goal or objective aligned with one or more Common Core content standards. If you are using curriculum materials, read and understand the learning goal(s) and objective(s) for the day's lesson as presented in the materials. As you consider the strengths and needs of the students in your classroom, you may feel that the curricular goal is not productive for some students, because it is either too easy or too hard. Think of ways to extend or modify the learning goal so that all students are working on related ideas but using approaches and strategies that are appropriate and accessible for them—allowing each student to engage in the mathematical practices by working on a "just right" goal. Number choice will help with this planning because multiple numbers let a range of students work on the same or similar goal(s) in ways that are most productive for each one. We focus in particular on the first two

phases of the lesson—number work and problem posing—as these are the phases in which planning with number choice is most prominent. We also highlight the Standards for Mathematical Practice that are best supported in each stage, recognizing that students may be engaging in multiple and overlapping mathematical practices during all four phases of this type of lesson.

Number Work

Most lessons start with a warm-up before the main task or presentation. We call this phase *number work* because we use it to focus students' attention on numbers: the meaning of numbers and the relationships and patterns in our number system. The number work phase of the lesson structure can include tasks such as true/false number sentences, open-ended number sentences, counting sequences, number of the day, or model representations such as dot cards, ten-frames, and base-ten blocks. It also might include an opening routine or warm-up from curriculum materials such as "Math Messages" (*Everyday Mathematics*, UCSMP 2007) or "Ten-Minute Math" (*Investigations*, TERC 2008).

Number work is usually done in a whole-group setting with the teacher guiding the students. Students might also have math partners with whom they engage in short bursts of discussion as questions or problems are posed. Students might use dry-erase boards or number-work notebooks to record their thinking. The number-work phase of the lesson typically lasts ten to fifteen minutes, although it might occasionally be extended.

In terms of number choice during the number-work phase of the lesson, it is important to consider how the numbers selected for the number work support the overall learning goal of the main lesson. The numbers chosen for number work should underpin or scaffold students' work with the multiple number choices that will be posed in the next phase of the lesson. For example, if one of the number choices planned for the problem-posing phase of the lesson is (10, 60), and the equation that matches the story problem is 10 + _____ = 60, then the number work might include a counting sequence by 10s from 10 to 60 (or beyond). Questions such as "How many tens are there from 10 to 60?" and "How would you represent that many tens with a number?" could be asked.

Number work helps students engage in multiple mathematical practices, including the following:

Standard for Mathematical Practice 7: *Look for and make use of structure.* Recognizing patterns and structures in mathematics is the essence of this standard. To support this practice, choose numbers during number work that reflect important

relationships and properties of numbers, as in the scenario below that focuses on the commutative property.

Standard for Mathematical Practice 8: *Look for and express regularity in repeated reasoning.* Number work often includes a string of related number sentences or counting sequences that help children identify patterns and regularity across the problems such as 10, 20, 30, and so on, or 13, 23, 33, and so forth.

Standard for Mathematical Practice 3: *Construct viable arguments and critique the reasoning of others.* In the example below, look at the questions that Jenny asks to encourage constructing and critiquing arguments. In the number work phase of the lesson, teachers often scaffold students in "build[ing] a logical progression of statements" (NGA Center and CCSSO 2010) through their choice of a series of number sentences that build toward a more general understanding of a particular number relationship or property. In our counting sequence examples above, a student might notice that when counting by tens, the ones place does not change while the tens place does.

Classroom Scenario: Phase 1

In Jenny Johnson's first-grade classroom, students work on developing the concept of the commutative property. Mrs. Johnson often begins the class with number work. Based on many similar discussions that Jenny has led, a typical dialogue that could occur in her class is presented below:

Mrs. Johnson:	Today we are going to look at an equation: 4 + 1 = 1 + 4. Is that statement true or false? Tell us why you believe the way you do.
Fred:	I believe it is true, because you just switch the numbers around.
Alice:	It's true because if you have 4 blue cubes and 1 white cube you have 5 and if you have 1 blue cube and 4 white cubes you still have 5.
Jack:	I know 4 + 1 is 5 and 1 + 4 = 5.
Mrs. Johnson:	It seems like everyone agrees that this statement is true. I will pose another problem that I'd like you to think about: 5 + 7 = 7 + _____. Think about what number should replace the blank; then you'll have a chance to talk to your math partner to see if you both agree.
Mrs. Johnson:	Is there a partner pair that would like to share?
Nora and Clare:	We think it should be 5. On the problem before, we just switched the numbers, so if you switch the 7, you have to switch the 5.
Grace and Grant:	We disagree with Nora and Clare. We think it should be 12.

Teacher Talk

Mrs. Johnson frequently does number work before introducing a problem to her students so they can recall and use some of that information to help them solve the posed problem. This is an example of engaging students in Standard for Mathematical Practice 7 by using what they know to help solve what they do not know.

Mrs. Johnson: Can you tell us a little bit more?

Grace and Grant: We added 5 and 7 and it was 12.

Fred and Alice: But what about the other 7?

Grace and Grant: We just left it there.

Mrs. Johnson: Are there any other groups that agree with Grace and Grant?

Alice and Fred: We don't, but here's how we thought about it. We added 5 and 7 and got 12. So we knew we had to make 12 on the other side of the equal sign. So we started at 7 and added up until we got to 12. We kept track on our fingers and found out it was 5.

Problem Posing

Problem posing is the lesson phase in which the teacher presents and unpacks the problem, and asks students to be thinking about it and developing a plan for solving it. As part of unpacking, the teacher should engage students in discussion in order to support and informally assess students' understanding of the problem itself. Problem posing is also the phase in which the use of number choice is most prominent. (Strategies for and examples of using number choice to address the Common Core content standards through problem posing are the central subjects of the subsequent chapters.) In general, the strategy for number choice in this phase of the lesson is to provide a range of number choices that 1) provide access to the learning goal at an appropriately challenging level for all students in the classroom, 2) build or relate to one another in ways that reflect a learning progression (Land and Drake 2014), and 3) will support a whole-class discussion later in the lesson. When using curriculum materials, planning for this phase will often involve choosing only one or two of the multiple problems that are provided in the curricular lesson. After choosing one or two problems, make those problems the focus of the lesson and ask students to develop their own strategies for solving the problem before (or instead of) showing them a procedure for solving. In other words, starting the lesson with problem solving as the main focus is more productive than starting the lesson with practicing a procedure and ending with problem solving. Most textbook problems will also require the addition of multiple number choices to the single number choice typically provided. With multiple, strategically chosen number choices, this phase of the lesson helps students prepare to solve the problem with numbers that are "just right" for them.

By helping students access the problem and prepare to solve it, problem posing facilitates students' engagement in these mathematical practice standards:

Standard for Mathematical Practice 1: *Make sense of problems and persevere in solving them.* According to NCTM, "problem solving means engaging in a task for which the solution is not known in advance" (2000, p. 52). Therefore, if a student knows a solution or is given a solution, they are not engaged in problem solving. Persevere means to persist or keep trying. In any given problem with a single number choice, there will be students who struggle in getting started as well as students who already have a solution. The former cannot engage and persevere in problem solving if the task is beyond their understanding, and the latter are not problem solving if they already have a known solution. Number choice and the multiple number choice lesson structure can help address each of these situations, and therefore, foster student engagement in the first Standard for Mathematical Practice: Easier number choices provide the means for students to access the problem without lowering the level of cognitive demand, and larger or more complex number choices supply opportunities for students to use invented algorithms, to think relationally, and to make connections. In addition, unpacking the problem gives students access to it in ways that help them make sense of the problem and, therefore, allows them to persevere in trying to solve it.

Unpacking a problem can also help students engage in Standard for Mathematical Practice 7 (look for and make use of structure) as they look within and across the multiple number choices.

Classroom Scenario: Phase 2

Consider how the following unpacking discussion would help students access the posed problem as they prepare to solve it:

Mrs. Johnson: Today for our math time I would like you to think about the following problem:

Ellie received _____ cards on Monday. On Tuesday, she received _____ more cards. How many cards did Ellie receive?

(3, 10)   (5, 23)   (10, 32)

Mrs. Johnson:	What is happening in our story?
Jamal:	Ellie is getting cards; maybe it's her birthday.
Mrs. Johnson:	Did she receive the cards all on one day?
Jamal:	No, she got some on Monday and some on Tuesday.
Mrs. Johnson:	Do we know how many she got each day?
Jasmine:	Yes, it tells us she got 3 on Monday and 10 on Tuesday.
Mrs. Johnson:	What is the action of this problem?
Alice:	It's a putting together problem.
Mrs. Johnson:	So you know what you need to do to solve the problem.

Student Work Time

Students may work in several ways to solve a problem: individually, with a partner, or even in a small group with teacher support. Students will be looking at the number choices to find relationships, and will then select a strategy that best matches their mathematical level and the given number choices. As students complete their work, they may conference with a peer to discuss the different strategies that they have used for the various number choices. The teacher has the flexibility during this stage to have conversations with students about their work and how they are thinking about solving the problem. The teacher should also be anticipating whose math work would meet a curricular goal or standard and should be highlighted during sharing time. Often, curriculum materials provide guidance or suggestions as to particular strategies and errors to look for during this part of the lesson and specific questions to ask students.

It is during work time that students may have opportunities to engage in all of the Standards for Mathematical Practice. In particular, they should be focused on:

Standard for Mathematical Practice 1: Make sense of problems and persevere in solving them.

Standard for Mathematical Practice 2: Reason abstractly and quantitatively.

Standard for Mathematical Practice 4: Model with mathematics.

Standard for Mathematical Practice 7: Look for and make use of structure.

For example, consider the following problem (also presented in chapter 4, page 32) and how the multiple number choices in the problem could support students in engaging in Standard for Mathematical Practice 2:

• Sam had _____ fishbowls. He had _____ fish in each bowl. How many fish did he have?

(5, 4)   (10, 4)   (7, 20)   (8, 20)

In this problem, we include number choices for which there are relationships among the factors so that a teacher can ask questions that prompt students to think about the number of fish (object) in each bowl (group) and how changing one of those (either the object or group number) affects the product. We suggest asking questions such as "How did the number of fishbowls change? What did that change do to the number of total fish?" By posing this problem and asking these types of questions, we are asking students to reason abstractly and quantitatively. "Mathematically proficient students make sense of quantities and their relationships in problem situations" (NGA Center and CCSSO 2010, p. 6). With these number choices, students can make sense of the number of fish in each bowl and the relationship between two sets of bowls (e.g., the second number choice is double the first). This Standard for Mathematical Practice also focuses on students' ability to decontextualize a situation by representing it symbolically (e.g., with an equation), so, when it is appropriate, we encourage students to write equations to represent their solutions across the number choices.

Further, this practice "entails habits of creating a coherent representation of the problem at hand; considering the units involved; attending to the meaning of quantities, not just how to compute them; and knowing and flexibly using different properties of operations and objects" (NGA Center and CCSSO 2010, p. 6). *Creating a coherent representation* of the problem includes direct modeling, drawing pictures, or using an equation, which meets Standard 2 (as well as Standard for Mathematical Practice 4). Asking the suggested questions about the fishbowls engages students in *attending to the meaning of the quantities*, not just computing them. The first two number choices, (5, 4) and (10, 4), provide an opportunity for students to *apply properties of operations*, because $10 \times 4 = 2 \times (5 \times 4)$.

As another instance of student engagement in the Standards for Mathematical Practice during student work time, we present two examples of student work on the following page. These cases highlight students modeling with mathematics (Standard for Mathematical Practice 4) and making use of the structure of numbers (Standard for Mathematical Practice 7).

In Kennady's work, she made multiple models (with a scaffold) of the possible number of yellow and red apples, each with a picture and equation. She also depicted the relationship between the two models accurately (e.g., 8 red apples and 2 yellow apples are also modeled by the equation 8 + 2). Grace's solution for the Mice in the Cage problem is different in that she modeled the situation with equations

only, but was more systematic in generating the equations by using the commutative property. Using the commutative property in this way is an example of *looking for and making use of structure* (Standard for Mathematical Practice 7).

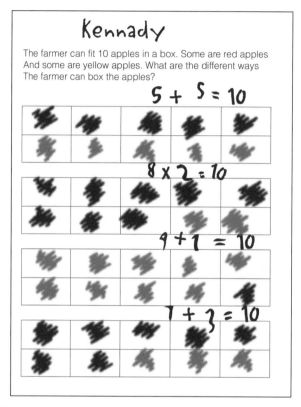

Fig. 3.1. Kennady made multiple models with a picture and equation for each.

Fig. 3.2. Grace modeled her problem only with equations.

Sharing Session

The teacher uses this last phase of the lesson, the sharing session, as a vehicle to highlight the mathematical content that was demonstrated in students' written work and oral explanations. Using student work as the basis of the sharing session is a standard technique for many teachers. The teacher's role is to connect strategy levels across students. Oftentimes, the teacher might present the work in a purposeful progression while selected students are asked to share their work as part of the progression. In this way, students can see what the next step for them might be. Having students recopy their work on a whiteboard or chart tablet or displaying student work with a document camera are just a few ways to engage the students and use their work in this phase. Questioning students, especially those who are reticent, is a potent tool for supporting students in making connections across ideas and strategies. Again, curriculum materials may provide useful questions for this fourth phase of the lesson.

The following practices are emphasized during a whole-class discussion when that dialogue is focused on the learning goal and student strategies across multiple number choices:

Standard for Mathematical Practice 1: Make sense of problems and persevere in solving them.

Standard for Mathematical Practice 2: Reason abstractly and quantitatively.

Standard for Mathematical Practice 3: Construct viable arguments and critique the reasoning of others.

Standard for Mathematical Practice 7: Look for and make use of structure.

Standard for Mathematical Practice 8: Look for and express regularity in repeated reasoning.

Returning to the Birthday Cards problem, consider the following first-grade sharing session (again, constructed on the basis of many similar discussions in Jenny Johnson's classroom) and how it illustrates students' engagement in these Standards for Mathematical Practice, particularly "construct viable arguments and critique the reasoning of others."

- Ellie received _____ cards on Monday. On Tuesday, she received _____ more cards. How many cards did Ellie receive?

                    (3, 10)   (5, 23)   (10, 32)

Classroom Scenario: Phase 4

Mrs. Johnson: I saw lots of different ways to solve this problem. I would like Grace to start us off on our sharing. (Mrs. Johnson had made notes of which students she would like to have share on the basis of what she observed the students doing during the work time. Her selections are purposeful and not haphazard. She hopes to highlight some strategies that she would like to see more of her students attempt.)

Grace: I took out 3 cubes, and then I took a ten stick and counted each one of these [3] and all of the cubes on the ten stick like 1, 2, 3, 4, . . . 13. (Note: Grace counted each individual cube on the ten stick.)

Clare: I did it kind of like Grace, but I knew instead of the problem being 3 + 10, I could switch it and make it be 10 + 3. Then I started at 10 and counted on 11, 12, 13.

Mrs. Johnson: Did anyone else do that as well? Grace, did you understand what Clare did? Could you explain it in your own words? Do you have any questions for her? Clare, can you tell us why you started with the 10 instead of the 3?

Clare: It was easier, and I didn't have to count as much.

Mrs. Johnson: Now, I would like Fred to share.

Fred: I added 5 and 23. I did it like Clare. I started at 23 and counted on 24, 25, 26, 27, and 28.

Mrs. Johnson: Can you tell us why you chose to start at 23 and not at 5?

Fred: It was easier for me to do that. It takes less counting.

In the subsequent chapters, we present examples of problems and number choices to address the content standards of the Common Core while engaging students in the Standards for Mathematical Practice.

Chapter 4

Using Number Choice to Meet Mathematical Content Standards

Examine this student work from the Pennies problem (introduced in chapter 1, page 1). What do you notice?

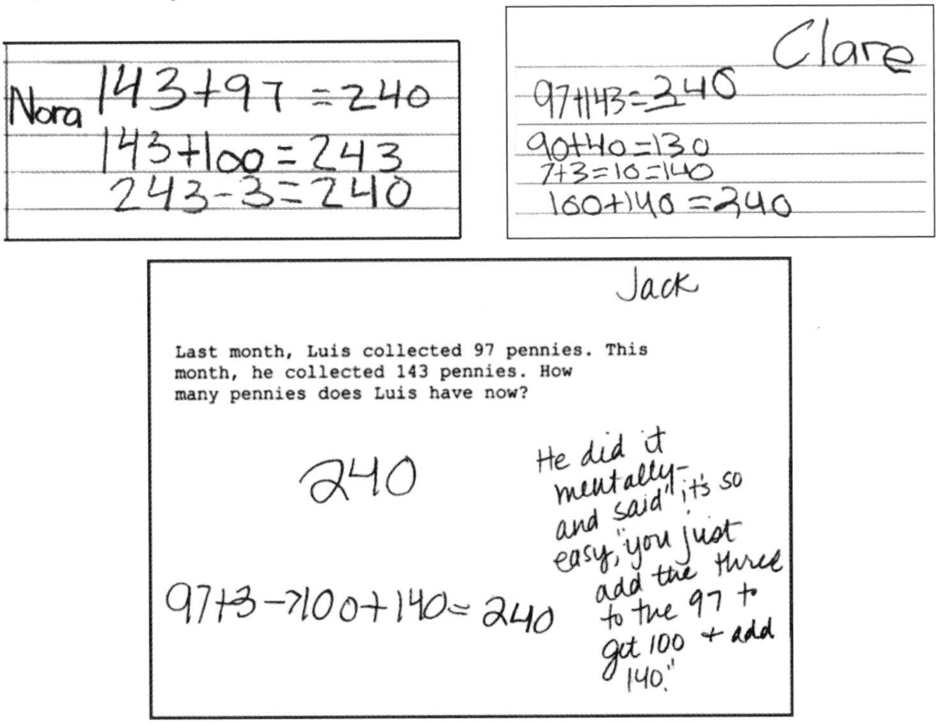

Nora
$$143 + 97 = 240$$
$$143 + 100 = 243$$
$$243 - 3 = 240$$

Clare
$$97 + 143 = 240$$
$$90 + 40 = 130$$
$$7 + 3 = 10 = 140$$
$$100 + 140 = 240$$

Jack

Last month, Luis collected 97 pennies. This month, he collected 143 pennies. How many pennies does Luis have now?

$$240$$

$$97 + 3 \rightarrow 100 + 140 = 240$$

He did it mentally and said "it's so easy, you just add the three to the 97 to get 100 + add 140."

Fig. 4.1. Student work for the Pennies problem

We noted that Nora switched the order in which she added the numbers, demonstrating knowledge of the commutative property of addition. The Common Core Standards 1.OA.B.3, 1.NBT.C.4, 1.NBT.C6, and 2.NBT.B5 are repeatedly addressed in first- and second-grade classrooms to set the foundation of knowledge for using the commutative property with questions like the Pennies problem and its larger-number choices. Evidence of place-value and base-ten understanding (standards K.NBT.1, 1.NBT.2, 2.NBT.1, and several others) is shown in all the strategies. For example, Clare ignored the hundreds place while adding the tens and ones; then added the 100 back in her last equation. Jack demonstrated knowledge of base ten by incrementing (adding a number in parts). Furthermore, Jack composed a 10 with the 7 and 3 as well as decomposed 143 into 140 and 3 to arrive at the answer. Clare decomposed 97 into 90 and 7 and 143 into 100, 40, and 3. Composing and decomposing numbers are addressed in standards, K.OA.3, K.NBT.1, 1.NBT.4, and 2.NBT.7.

These types of strategies are the result of numerous opportunities to work with numbers and make sense of numbers. Students need many prior experiences with numbers before they are able to develop this type of content knowledge—the kind of knowledge called for in the Common Core Standards for the early grades (K–2). Nora, Clare, and Jack's classroom teachers all had very similar math class structures consisting of blocks of number work, problem solving, and student sharing. Choosing the right numbers and helping students make connections between the number work and the number choices used in problem solving are critical facets of the teacher's role.

Addressing Content with Number Choices

Here, we discuss how number choices can be created to address different content throughout the Common Core. In chapter 1, the number choices for The Fishbowl problem were created to support standards 2.OA and 3.OA. Below, we rewrite the Fishbowl problem to support several other standards by formulating appropriate number choices for each standard.

K.CC.1: Count to 100 by ones and by tens.

- Sam had _____ fishbowls. He had _____ fish in each bowl. How many fish did he have?

A	(5, 1)	(10, 1)	(17, 1)	(30, 1)
B	(2, 10)	(4, 10)	(5, 10)	(10, 10)

In the problem above, students can use counting by ones and tens to solve the problem for the various number choices. When students solve for the numbers in the first row, you want them to recognize that if there is one fish in each bowl, the answer is the number of bowls. This type of number choice helps students develop knowledge about conservation of number and multiplicative relationships and properties. A question you might ask would be, "What do you notice about the number of fish and the number of fishbowls?"

For the second row of numbers, we recognize that students may not use counting by tens initially because they may use objects or draw pictures to model the problem. Students can be prompted to count by tens to solve these number choices either by doing an opening number routine where students are counting by tens and/or by asking students who did count by ten to share their strategy with the class.

The Common Core does not mention other numbers that children need to count by in kindergarten or first grade, but we feel it is important for students to also have experiences counting by twos and fives in these grades. Therefore, we include the following revision of the Fishbowl problem to demonstrate how to meet that need:

- Sam had _____ fishbowls. He had _____ fish in each bowl. How many fish did he have?

A	(5, 2)	(10, 2)	(20, 2)	(25, 2)
B	(2, 5)	(4, 5)	(5, 5)	(10, 5)

Initially, many students will count by ones with these number choices, which is fine, because that counting method still meets the standard. For the second row of number choices, students may recognize that two 5s makes a 10 and use that knowledge to solve for some of the number choices and meet the original standard.

You may also want students not to start at zero each time they solve the problem for a new number choice, but, instead, count on from a previous number choice. To do that, you might ask a student, "How can you use what you know from this number choice (e.g., (5, 5)) to solve for this number choice (e.g., (10, 5))?" or "Do you need to start at zero for (10, 5); can you start where you left off with (5, 5)?"

2.OA: Work with equal groups of objects to gain foundations for multiplication.

- Sam had _____ fishbowls. He had _____ fish in each bowl. How many fish did he have?

(5, 4)   (7, 5)   (5, 20)   (8, 11)

For this content standard, you could use a variety of number choices depending on the understandings of your students. Multiplication problems can be quite easy for students if the problems are written in contexts with which they are familiar. Other situations often used are packs of gum or people riding in a car. Also included in the 2.OA standard is the following, "Use addition to find the total number of objects arranged in rectangular arrays with up to 5 rows and up to 5 columns; write an equation to express the total as a sum of equal addends" (Standard 2.OA.4; NGA Center and CCSSO 2010). The circumstances of the Fishbowl problem do not fit well with this section of the standard, but a context such as rows in a garden, candy bars divided into portions, or seats in an auditorium would all work well.

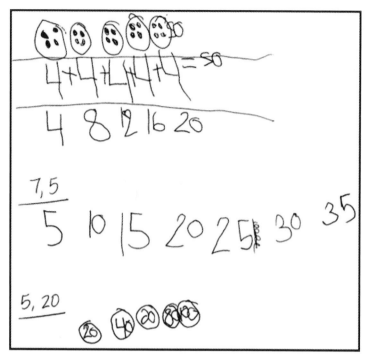

Fig. 4.2. Grace's work for the Fishbowl problem

Because multiplication is accessible to young students, multiplication problems can be posed as early as kindergarten. Multiplication problems are also powerful content resources because they provide opportunities to work on several standards within one problem. As an example, we present work from Grace, a kindergartener. As you can see in Grace's work, the Fishbowl problem was easily accessible to her. In fact, she represented the first number choice in three different ways. With her last two number choices, Grace was able to use skip counting and no longer used individual representations of the fish in each bowl. Her change in strategy might be because the number choices are easier for her (fives and twenties instead of fours) or because she no longer needed to direct model after doing so with the first number choice. Grace's work demonstrates understanding in standard 2.OA.2 as well as three others:

1. K.CC.3: Represent a number of objects with a written numeral 0–20.

2. K.OA.1: Represent addition.

3. 2.NBT.2: Skip-count by 5s, 10s, and 100s.

Mathematical Practices Connections

Mathematical Practice 4: *Model with mathematics.* Grace used this practice by modeling the problem situation with each number choice (see student work). You can have all students utilize this practice by asking them to model, in some way, the problem situation.

Mathematical Practice 7: *Look for and make use of structure.* Grace utilized this practice by coming to know that she could solve multiplication problems by using skip counting, and then further used that structure in subsequent number choices.

Fig. 4.3. The mathematical practices that Grace engaged

2.NBT.2: Count within 1000; skip-count by 5s, 10s, and 100s.

• Sam had _____ aquariums. He had _____ fish in each aquarium. How many fish did he have?

(5, 100)   (8, 100)   (10, 100)   (15, 100)

For this content standard, we chose to focus on the number 100 just as previous problems focused on 5s and 10s. We changed the context to aquariums because an aquarium is more appropriate for the number of fish posed. For more problems that meet the 2.NBT standards, see Appendix C (page 99).

3.OA.1: Interpret products of whole numbers, e.g., interpret 5 × 7 as the total number of objects in 5 groups of 7 objects each.

• Sam had _____ fishbowls. He had _____ fish in each bowl. How many fish did he have?

(5, 7)   (10, 7)   (7, 20)   (8, 20)

This standard is similar to the 2.OA standard, but it emphasizes interpreting the product of whole numbers. Therefore, we chose numbers where there were relationships among the factors so that a teacher could ask questions that prompted students to think about the number of fish (object) in each bowl (group) and how changing one of those factors (either the object or group number) affects the product. For example, a teacher might ask, "In the first two number choices, what stayed the same—the number of fish or the number of bowls?" and "How did the number of fishbowls change? What did that change do to the number of total fish?" For the second two number choices, a teacher might ask, "What changed in this case?" and follow up, "How does adding one more bowl of fish change the total number of fish?"

Mathematical Practices Connections

Mathematical Practice 2: *Reason abstractly and quantitatively.* By posing the Fishbowl problem for Standard 3.OA.1 and asking questions like those suggested, one is encouraging students to engage in this practice. "Mathematically proficient students make sense of quantities and their relationships in problem situations" (NGA and CCSSO 2010, p. 6). With these number choices, students can make sense of the number of fish in each bowl and the relationship between two sets of bowls; for example, the second number choice is double the first. This practice also focuses on the ability to decontextualize a situation by representing it symbolically, for instance, with an equation. Equations that would represent using the answer from the first number choice (5, 7) to help solve for the second (10, 7) are 5 x 7 = 35 and 35 x 2 = 70.

In addition, Practice 2 "entails habits of creating a coherent representation of the problem at hand; considering the units involved; attending to the meaning of quantities, not just how to compute them; and knowing and flexibly using different properties of operations and objects" (NGA and CCSSO 2010, p. 6). Direct modeling, drawing a picture, or writing an equation all meet the requirement of *creating a coherent representation* of the problem. Asking the suggested questions directs students to *attend to the meaning of the quantities*, not just compute them. The first two number choices ((5, 7) and (10, 7)) provide an opportunity for students to *use different properties of operations* because it is easier to use the commutative property to find the answer.

Fig. 4.4. The mathematical practice engaged in the Fishbowl problem

A different mathematical situation for this content standard is to provide the total number of objects (fish), and then ask students to determine how many individual groupings are possible:

- Sam had _____ total fish. Sam wants to divide his fish equally into different bowls. How many different combinations of fish in fishbowls can Sam make?

(16)   (17)   (24)   (36)   (60)

We have included number choices here for which there are many different combinations, as well as one choice (17) for which there are only two combinations. This type of problem also serves as a precursory experience for the fourth-grade standard 4.OA.4: *Find all factor pairs for a whole number in the range 1–100. Recognize that a whole number is a multiple of each of its factors. Determine whether a given whole number in the range 1–100 is a multiple of a given one-digit number. Determine whether a given whole number in the range 1–100 is prime or composite.*

3.OA.5: Understand properties of multiplication and the relationship between multiplication and division.

- Sam had _____ fishbowls. He had _____ fish in each bowl. How many fish did he have?

    (6, 7)   (7, 6)   (12, 7)   (7, 12)

This standard refers to the commutative, associative, and distributive properties of multiplication. In the problem above, we provided the same number choice for the first and second two pairs of numbers, simply switching the order. A teacher would want to follow up a problem like this with a question such as, "How can the total number of fish be the same in the first two cases?"

In the following problem, the associative property is addressed by adding a third factor to the number choices:

- At the pet store, there are _____ rows of aquariums. In each row, there are _____ aquariums. In each aquarium, there are _____ fish. How many fish are for sale at the pet store?

    (5, 7, 2)   (2, 7, 5)   (7, 2, 5)

A follow-up to this problem would be to reverse the order of the story using the same number choices and ask students what they notice about the answers for these two questions.

For problems addressing both the commutative and associative property, you might consider using contexts that lend themselves to arrays, for example, rows in a garden. Arrays provide a visual situation for students to draw, making it easier for some to understand and to prove the two properties. A figure can be drawn for 12 rows of 5 tomato plants and for 5 rows of 12 tomato plants. As students compare the two pictures, they may notice that by rotating the images, they can depict each tomato plant situation.

In the third problem below, we offer two columns of numbers that might help students understand the distributive property. We arranged the numbers in columns because it makes each set of numbers more distinct.

- Sam had _____ fishbowls. He had _____ fish in each bowl. How many fish did he have?

A	B
(2, 3)	(2, 8)
(10, 3)	(10, 8)
(12, 3)	(12, 8)

To help students understand the distributive property, a teacher wants students to notice that the sum of the products for the first two number choices in each column equals the product for the third number choice. To do that, a teacher might ask students to write a reflective piece on how the products are related. Another strategy would be to ask students to use the first two number choices to help with the third.

4.NF.4: Solve word problems involving multiplication of a fraction by a whole number, e.g., by using visual fraction models and equations to represent the problem.

- Sally is baking peanut butter cookies for a bake sale. She needs _____ cups of sugar for one batch. If she bakes _____ batches, how much sugar will she need?

$$(1/2, 4) \quad (1/2, 20) \quad (1/4, 8) \quad (1/4, 9)$$

For this standard, we needed to change the context of our problem. There are no fractional parts of a fishbowl or (we hope) no fractional parts of fish in the fishbowl. For this version of our problem, we changed the context to baking cookies.

We started with halves because halves are easy to work with, and then moved to fourths because of the relationship between ½ and ¼. You may have noticed that the whole number from the third choice was increased by one in the last number choice. Increasing the number by one was intended to provide a scaffold into a number choice that would not result in a whole number when multiplied.

The Pennies and Toy Cars Problems

In this section, we present the Pennies problem with a variety of number choices to meet various content standards. For the problems formulated to meet kindergarten standards, we changed the context to toy cars with no reference to months because we thought it would be easier for kindergarten students to relate to and understand. The problem type (add to–result unknown) for both contexts is the same.

KCC: Count to tell the number of objects.

- Joey has _____ toy cars. His mom gave him _____ toy cars for his birthday. How many toy cars does Joey have now?

$$(2, 3) \qquad (4, 4) \qquad (8, 2) \qquad (5, 5)$$

While working toward this standard, the objective is to provide students with opportunities to count objects. Even if students cannot solve the problem correctly, they are still getting practice counting objects. Additionally, this problem would meet the following kindergarten standards:

KOA.1: Represent addition and subtraction with objects, fingers, mental images, drawings, sounds (such as claps), acting out situations, verbal explanations, expressions, or equations.

KOA.2: Solve addition and subtraction word problems, and add and subtract within 10, e.g., by using objects or drawings to represent the problem.

KOA.5: Fluently add and subtract within 5.

- Joey has _____ toy cars. His mom gave him _____ toy cars for his birthday. How many toy cars does Joey have now?

$$(0, 5) \qquad (1, 4) \qquad (2, 3) \qquad (3, 2) \qquad (4, 1) \qquad (5, 0)$$

To meet this standard, students will need several experiences with adding numbers up to 5. We included all the number choices that would have a sum of 5 in the hope students would begin to recognize all the combinations for 5. Of course, a teacher would want to include other number choices with sums within 5 in additional problems.

K.NBT.1: Compose and decompose numbers from 11 to 19 into ten ones and some further ones, e.g., by using objects or drawings, and record each composition or decomposition by a drawing or equation (e.g., 18 = 10 + 8); understand that these numbers are composed of ten ones and one, two, three, four, five, six, seven, eight, or nine ones.

- Joey has _____ toy cars. His mom gave him _____ toy cars for his birthday. How many toy cars does Joey have now?

$$(10, 1) \qquad (10, 3) \qquad (10, 8) \qquad (3, 10) \qquad (9, 10)$$

In this case, one number in each number choice is always 10. Keeping 10 as a constant means that the total number is composed of 10 toy cars plus some more. With multiple exposures to number choices that involve keeping 10 constant, students will see that with any single-digit number plus 10, the digit in the ones place of the sum is the same as the single-digit number. By reversing the position of the 10 in the last two number choices, we are promoting the use of the commutative property.

1.OA.1: Use addition and subtraction within 20 to solve word problems involving situations of adding to, taking from, putting together, taking apart, and comparing, with unknowns in all positions, e.g., by using objects, drawings, and equations with a symbol for the unknown number to represent the problem.

- Last month, Luis collected _____ pennies. This month, he collected _____ pennies. How many pennies does Luis have now?

$$(5, 3) \qquad (8, 8) \qquad (8, 9) \qquad (15, 3)$$

In this problem, we are meeting part of the standard with a putting together–result unknown situation. To fully meet the standard, other types of problems would need to be posed.

1.OA.2: Solve word problems that call for addition of three whole numbers whose sum is less than or equal to 20, e.g., by using objects, drawings, and equations with a symbol for the unknown number to represent the problem.

- Last month, Luis collected _____ pennies. This month, he collected _____ pennies and _____ nickels. How many coins does Luis have now?

$$(3, 7, 5) \qquad (6, 6, 2) \qquad (1, 8, 2) \qquad (4, 9, 3)$$

Some nickels are added to this situation to include a third whole number.

1.OA.3: Apply properties of operations as strategies to add and subtract. Examples: If 8 + 3 = 11 is known, then 3 + 8 = 11 is also known. (Commutative property of addition.) To add 2 + 6 + 4, the second two numbers can be added to make a ten, so 2 + 6 + 4 = 2 + 10 = 12. (Associative property of addition.)

- Last month, Luis collected _____ pennies. This month, he collected _____ pennies. How many pennies does Luis have now?

$$(7, 3) \qquad (7, 4) \qquad (4, 7)$$

Student understanding of the commutative and associative properties of addition is intrinsic to this standard. Therefore, we focused on providing an opportunity for students to notice that when adding the same numbers but in a different order, they arrive at the same answer. Students can come to know and understand this generalization (you can add numbers in any order) through multiple experiences with number choices like those above. During these encounters, a teacher might ask, "Does that always happen when adding?" (To work on applying properties as strategies, a teacher would approach number choices in a slightly different manner that will be discussed in chapter 5.) Next, we provide a problem for work with the associative property:

- Last month, Luis collected _____ pennies. This month, he collected _____ pennies. Next month, he hopes to collect _____ pennies. How many pennies will Luis have?

A	(7, 1, 9)	(9, 7, 1)	(9, 1, 7)
B	(6, 6, 2)	(2, 6, 6)	(6, 2, 6)

The first row of numbers promotes making tens, while the second promotes the use of doubles. After posing number choices like these, a teacher might ask, "Which of these number sets was easiest to solve? Why?" More problems and number choices for the 1.OA standards and associative property are in Appendix C (page 99).

1.NBT.4: Add within 100, including adding a two-digit number and a one-digit number, and adding a two-digit number and a multiple of 10, using concrete models or drawings and strategies based on place value, properties of operations, and/or the relationship between addition and subtraction; relate the strategy to a written method and explain the reasoning used. Understand that in adding two-digit numbers, one adds tens and tens, ones and ones; and sometimes it is necessary to compose a ten.

This standard is quite dense, but if we unpack it a little, we see that the main content of the standard is adding within 100, adding two-digit and one-digit numbers, and adding a multiple of 10 and a two-digit number. This standard additionally calls for students to understand that in adding two-digit numbers, an individual adds tens and tens, and ones and ones. The remainder of the standard is a focus on strategies and mathematical practices, so we will visit this standard again in chapter 5. Because we have addressed similar standards to adding within 100, such as adding within 20, and because adding two-digit and one-digit numbers is relatively self-explanatory, we focus our next two problems on adding multiples of ten and understanding place value.

Mathematical Practices Connections

Mathematical Practices 2 and 3: *Reason abstractly and quantitatively; construct viable arguments and critique the reasoning of others.* To relate the strategy to a written method and explain the reasoning used, standard 1.NBT.4 requires students to engage in these practices. A teacher could also highlight **Mathematical Practice 6,** *attend to precision,* by having students use precise mathematical language in their explanations.

Fig. 4.5. The mathematical practices that connect to the following two problems

- Last month, Luis collected _____ pennies. This month, he collected _____ pennies. How many pennies does Luis have now?

A	(10, 10)	(30, 10)	(30, 20)	(30, 40)
B	(17, 10)	(17, 40)	(37, 40)	(37, 70)
C	(10, 57)	(30, 57)	(20, 54)	(20, 64)

We provided three rows of numbers for this problem to illustrate several different number choice combinations and progressions. In the first row, we posed all multiples of tens as an entry point into this standard. The second row consists of adding a non-multiple of ten with a multiple of ten. Because it is a hard transition for some students to make, the last number choice in that row goes over the hundred to start laying the foundation for adding numbers that go over 100 in second grade. In the third row, we reversed the order of the multiple of ten and the non-multiple of ten.

For understanding that in adding two-digit numbers, one adds tens and tens, and ones and ones, we present the following problem:

- Last month, Luis collected _____ pennies. This month, he collected _____ pennies and _____ nickels. How many coins does Luis have now?

   (30, 10, 5)   (40, 30, 2)   (41, 40, 2)   (64, 20, 7)

In this problem, students are adding three numbers with a focus on adding some tens and then some ones to the first number in each set. As students solve the problem, the teacher may want to go around and ask questions such as, "How many tens did you add to the first number? How many ones? How did those additions change the number?"

2.NBT.1: Understand that the three digits of a three-digit number represent amounts of hundreds, tens, and ones; e.g., 706 equals 7 hundreds, 0 tens, and 6 ones.

- Last month, Luis collected _____ pennies. This month, he collected _____ pennies. Next month, Luis plans to collect _____ pennies. How many pennies will Luis have after next month?

    (100, 20, 6)   (300, 90, 7)   (400, 40, 1)   (500, 0, 6)

To meet this standard, we have numbers that involve composing an answer with hundreds, tens, and ones separately. As before, teachers would want to ask questions that call students' attention to the place-value concepts.

3.OA.8: Solve two-step word problems using the four operations. Represent these problems using equations with a letter standing for the unknown quantity. Assess the reasonableness of answers using mental computation and estimation strategies including rounding.

- Last month, Luis collected _____ pennies. This month, he collected _____ pennies. Next month, he plans to spend _____ of his pennies on a new toy. How much money will Luis have left?

    (60, 10, 20)   (203, 10, 200)   (645, 55, 310)   (950, 220, 500)

With these choices, we thought about a couple of things. First, we provided occasions to work with tens and making hundreds. Second, there are opportunities for students to think strategically about the order in which they add and subtract the numbers. For instance, in the second number choice, students could first subtract the 200 from 203 to get 3, and then add 10. To fully meet the above standard, a teacher would have to ask students to write an equation representing the problem that included a letter to stand for the unknown quantity; students would also have to assess the reasonableness of the answers.

Conclusion

To start selecting number choices to meet Common Core Standards, pick a standard that you want to work on. Next, choose a problem type that is appropriate for that standard, and then a problem context. Recognize that multiple problems with multiple number choices will need to be posed to meet any one standard. When considering a problem situation, think about topics that are relevant to your students' lives—either at home or at school. Once the problem is written, pick a few number choices that meet the standard. Realize that standards at a particular grade level are an end-of-year goal; therefore, you can start with easier number choices and work toward harder ones. Anticipate solution strategies that your students will use, and then monitor them. As you see how your students respond to the different numbers, you will continue to learn how to pose appropriate numbers to meet the range of learners in your classroom.

Chapter 5

Strategies

As children develop physically, they achieve many milestones, but we know that not every child is going to crawl, walk, talk, or ride a bike at the same time. However, too often in elementary school mathematics classrooms, the assumption is that all students progress at the same rate in the same amount of time. Conversely, research about mathematics learning (e.g., Carpenter et al. 1999) has shown that each child has his or her own mathematics story and each child takes a different path to achieving mathematical understanding. Traditionally, mathematics classrooms and textbooks provide only one path for students to travel when solving problems, but by including various types of strategies in its standards, the Common Core has now opened up several paths for students to achieve mathematical understanding. Consider the following kindergarten standard for number and operations in base ten, paying particular attention to how students are to meet the standard:

> K.NBT.1: Compose and decompose numbers from 11 to 19 into ten ones and some further ones, e.g., by using objects or drawings, and record each composition or decomposition by a drawing or equations (e.g., 18 = 10 + 8): understand that these numbers are composed of ten ones and one, two, three, four, five, six, seven, eight, or nine ones (NGA and CCSSO 2010, p. 12).

In first grade, students still meet standards using concrete models or drawings, but other, more sophisticated strategies, such as those based on place value, are included in standards like 1.NBT.4, which suggests many means by which students can meet the standard.

> 1.NBT.4: Add within 100, including adding a two-digit number and a one-digit number, and adding a two-digit number and a multiple of 10, using concrete models or drawings and strategies based on place value, properties of operations, and/or the relationship between addition and subtraction; relate the strategy to a written method and explain the reasoning used. Understand that in adding two-digit numbers, one adds tens and tens, ones and ones; and sometimes it is necessary to compose a ten (NGA and CCSSO 2010, p. 14).

While the Common Core has included many means for students to meet a standard, it is also clear in the Common Core that students need to progress in their use of

strategies in order to continue meeting standards. For example, in second grade, using concrete models and drawings no longer meets the standard 2.NBT.5. By the end of second grade, students should be using "strategies based on place value, properties of operations, and/or the relationship between addition and subtraction" (NGA and CCSSO 2010, p. 19).

By not prescribing particular strategies in the early grades (the standard algorithms for addition and subtraction do not appear until fourth grade), the Common Core provides opportunities for students to invent and use approaches that are mathematically transparent and still meet the standards. For example, examine Clare's strategies for subtraction:

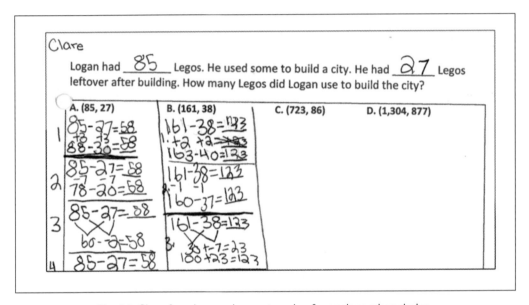

Fig. 5.1. Clare found several ways to solve for each number choice.

In her work, Clare found several ways to solve for each number choice. She demonstrated relational and creative thinking as she found numbers to add or subtract to the minuend and subtrahend to make the computation easier to do. For instance, in her first strategy for 85 − 27, she added 3 to both 85 and 27 because subtracting 30 from 88 is easier than subtracting the original numbers. Clare used this strategy of adding or subtracting numbers to the minuend and subtrahend several times. She also made use of negative numbers in her third strategy for 85 − 27.

It is up to us, as educators, to let students follow the path they need to take while still assisting them in their progression of strategies and allowing for creativity. In the remainder of this chapter, we demonstrate how to support children in their use of strategy with number choice.

Mathematicians make meaning in their world by setting up quantifiable and spatial relationships, by noticing patterns and transformations, by proving them as generalizations, and by searching for elegant solutions. They construct new mathematics to solve real problems or to explain or prove interesting patterns, relationships, or puzzles in mathematics itself. The renowned mathematician, David Hilbert, once commented that he liked to prove things in at least three or four different ways, because by doing so he better understood the relationships involved. At the heart of mathematics is the process of setting up relationships and trying to prove these relationships, mathematically, in order to communicate them to others. Creativity is at the core of what mathematicians do (Fosnot and Dolk 2001b, p. 4).

K.OA.2: Solve addition and subtraction word problems, and add and subtract within 10, e.g., by using objects or drawings to represent the problem.

Of all the problem types listed in the Common Core, there are at least four basic kinds that a teacher would want to pose in kindergarten (add to–result unknown, take from–result unknown, put together/take apart–total unknown, put together/take apart–part unknown) to achieve standard K.OA.2. There are a total of twelve addition/subtraction problem types, but these four are the easiest for students to solve. However, we know several teachers who also pose multiplication problems to kindergarten students because multiplication addresses learning goals related to addition, such as repeated addition and doubling. Students will likely use either direct modeling (representing both numbers and following the action of the problem) with manipulatives, drawings, and/or fingers, or counting strategies such as counting on or counting back to solve these problems (Carpenter et al. 1999). For Standard K.OA.2, we provide several problems with number choices that meet the standard. Because this standard suggests that students should represent these problems, students can meet the standard by direct modeling the problem, but it is important to transition students from direct modeling to counting strategies to prepare them for later standards that stress counting strategies, such as 1.OA.6 (page 52). Counting methods are the central avenues by which children move into strategies based on place value, which are emphasized in first grade as well as in the later grades.

Teacher Talk

We want students to use counting strategies to transition to using known facts and invented algorithms, but we do not want them to use counting strategies for long. In other words, students need to count in order to transition, but the time spent on counting should be limited.

Add To–Result Unknown Problem

- Ellie received _____ cards on Monday. On Tuesday, she received _____ more cards. How many cards did Ellie receive?

    (7, 2)   (1, 8)   (6, 4)

Take From–Result Unknown Problem

- Sam baked _____ cookies. He ate _____ of them. How many cookies does he have left?

    (8, 2)   (7, 3)   (10, 4)

Put Together/Take Apart–Total Unknown Problem

- There were _____ pink candles and _____ blue candles on the birthday cake. How many candles were there on the cake?

    (4, 2)   (8, 2)   (2, 10)

Put Together/Take Apart–Part Unknown Problem

- There were _____ balls in the playground bucket. _____ balls were red. The rest were blue. How many blue balls were there?

    (5, 3)   (2, 8)   (5, 9)   (7, 3)

As you can see, all the number choices are within 10. Students need to spend significant time solving problems that meet this standard to develop their problem-solving skills as well as number sense.

K.OA.3: Decompose numbers less than or equal to 10 into pairs in more than one way, e.g., by using objects or drawings, and record each decomposition by a drawing or equation (e.g., 5 = 2 + 3 and 5 = 4 + 1).

Put Together/Take Apart–Both Addends Unknown Problem

- There were _____ mice in the cage. Some of the mice were white and some were brown. How many of each color of mice could there be?

    (5)   (6)   (8)   (10)

Students might come up with one, multiple, or all the possible solutions for this type of problem. When students find all the possible solutions, they most likely will do so arbitrarily at first.

Teacher Talk

Using a ten-frame helps students organize their thinking and their work when operating with the number 10. An example is given in Appendix A, which can also be downloaded as a blackline master at NCTM's More4U site (nctm.org/more4u).

Kennady's work in chapter 3 (page 28) illustrates this type of rudimentary reasoning in which children develop multiple solutions without following a pattern. In other words, they at first do not have an organized way of finding the combinations, like systematically listing them all or making use of the commutative property, but they may develop a method after repeated experiences with this problem type. (Grace's work (chapter 3, page 28) is an example of how she recognized and used the commutative property.) Interestingly with these problems, the number of solutions is always one more than the number choice. Thus, for five mice, there are six possible solutions. If students notice that generalization, they are engaging in Mathematical Practice 7: Look for and make use of structure.

1.OA.3: Apply properties of operations as strategies to add and subtract.

In chapter 4 we focused on using number choices to help students notice and understand properties. Here, we concentrate on using properties as strategies, as the standard states, by not providing both or all addends in our number choices (e.g., 7 + 3 and 3 + 7). In the following problems, we instead provide situations where applying a property makes the problem easier to solve:

- Ellie received _____ cards on Monday. On Tuesday, she received _____ more cards. How many cards did Ellie receive?

$$(3, 10) \qquad (5, 23) \qquad (10, 32)$$

In each of these cases, it is easier to add on to the second number rather than the first, prompting students to use the commutative property to reverse the numbers in the problem. Some students, however, may still want to use a different strategy, for example, counting on from the first number, rather than use the commutative property. For these students, you could pose two number choices that switch the position of the numbers, for instance, (10, 32) and (32, 10), and ask how these two

number choices are the same and how they are different, which might then lead to the question of which number set was easier to solve. Another number choice set you could pose would be a high two-digit number paired with a low single-digit number, such as (2, 82), which makes it difficult to add from the first number, but easy to add from the second number. When students count on from the second number, they may explain that they "put the larger number first" and not verbalize that they are using the commutative property. It is up to the teacher to make the connection between counting up from the larger number and the commutative property. It is also up to the teacher to remind students of their "new" strategy on subsequent days by saying something like, "Yesterday, you started counting on from the second number. Why did that work for you? Could you do that again today?"

The associative property states that three or more numbers can be added in any order with the same result. Students should have numerous experiences using the commutative property before they are asked to apply the associative property. If you notice that students are grasping the commutative property, you can prompt them to begin thinking about the associative property by asking, "When adding two numbers, we have gotten the same answer no matter in which order we add them. Do you think that would work with three numbers?" In the next problem, we want students to apply the associative property.

- There were _____ pink candles, _____ blue candles, and _____ yellow candles on the birthday cake. How many candles were there on the cake?

$$(5, 2, 5) \quad (8, 7, 2) \quad (50, 6, 50) \quad (9, 18, 12) \quad (19, 34, 41)$$

Our number choices should prompt students to add the numbers in a different order than that given, and the choices become progressively more difficult. In the first two number choices, students could find two numbers that equal 10, which then makes it easy to add the third number. The third number choice includes a multiple of 10, but 50 + 50 is also accessible to most first graders as is adding 6 to 100. In the last two number choices, students could use making a ten with two-digit numbers first, and then add the third number, which can be a single- or double-digit number.

1.OA.5: Relate counting to addition and subtraction (e.g., by counting on 2 to add 2).

This standard is about counting on or back from a number and relating that action to an operation. Therefore, you would want students to hold the first number in their head and either add or subtract the second number instead of counting all the objects. Similar to the standard above, you can prompt students to count on by making it difficult for them to count all. To make counting all difficult, simply pose larger numbers as we do in the following problem:

- Ellie received _____ cards on Monday. On Tuesday, she received _____ more cards. How many cards did Ellie receive?

(22, 2)   (67, 3)   (54, 4)

To completely meet the standard, students would also need to relate their counting strategy to addition by writing an equation.

To further meet this standard, you could include additional number choices within a problem:

- I went outside and saw some insects. I saw _____ ladybugs, _____ moths, _____ mosquitos, and _____ praying mantis. How many insects did I see?

(8, 2, 2, 2)   (10, 5, 5, 5)   (70, 10, 10, 10)   (35, 5, 5, 5)   (41, 2, 2, 2)

As you can see, we have number choices that involve counting on by twos, fives, and tens. These are the easiest numbers for first graders to use and relate to. Additionally, the number choices involve starting from even and odd numbers when counting on by twos. When counting on by fives and tens, we have start numbers that are multiples of five and ten. You could also have a start number, such as 42, with the remaining numbers as 10, 10, and 10 to help the students learn how to count from a non-decade number.

Teacher Talk

When children are posed problems with multiple addends and minuends, it is important for teachers to model correct mathematical notation. Children will often record their thinking as $10 - 2 = 8 - 2 = 6 - 2 = 4$, but teachers should record any mathematical notation using proper form. In this case, it could look like—

$$10 - 2 = 8$$
$$8 - 2 = 6$$
$$6 - 2 = 4$$
$$\text{or } 10 - 2 \rightarrow 8 - 2 \rightarrow 6 - 2 \rightarrow 4$$

For counting back, we provide the following problem and number choices:

- Sallie the shark had _____ teeth. She lost _____ teeth on Monday, _____ teeth on Tuesday, and _____ teeth on Wednesday. How many teeth does she have left?

(10, 2, 2, 2)   (17, 2, 2, 2)   (30, 5, 5, 5)   (67, 10, 10, 10)   (101, 2, 2, 2)

Again, we chose the numbers 2, 5, and 10 to count back by. In this type of problem, you want students to make the connection between counting back and subtraction.

1.OA.6: Add and subtract within 20, demonstrating fluency for addition and subtraction within 10. Use strategies such as counting on; making ten (e.g., 8 + 6 = 8 + 2 + 4 = 10 + 4 = 14); decomposing a number leading to a ten (e.g., 13 – 4 = 13 – 3 – 1 = 10 – 1 = 9); using the relationship between addition and subtraction (e.g., knowing that 8 + 4 = 12, one knows 12 – 8 = 4); and creating equivalent but easier or known sums (e.g., adding 6 + 7 by creating the known equivalent 6 + 6 + 1 = 12 + 1 = 13).

This standard calls for students to demonstrate fluency. According to *Principles and Standards for School Mathematics* (NCTM 2000), a student is fluent if "they demonstrate *flexibility* in the computational methods they choose, *understand* and explain these methods, and produce accurate answers *efficiently*" (p. 152, italics added). Fluency does not only mean solving problems quickly, as the term is often interpreted. As you can see, standard 1.OA.6 mentions several different types of strategies students might use. In the following problems, we provide number choices to help students notice and use the various strategies listed in the standard and to scaffold their progress from one method to another.

Making Ten

• Ellie received _____ cards on Monday. On Tuesday, she received _____ more cards. How many cards did Ellie receive?

$$(6, 4) \quad (6, 5) \quad (16, 4) \quad (6, 8)$$

The first number choice makes a ten, which students could then apply to the second choice. Students might notice that they could take 4 from the 5 to make a ten with the 6, and then one more makes 11. Or, they might notice that 5 is one more than 4, so the answer has to be one more than 10. The third number choice involves making a 10 with the 6 and 4 and adding it to another 10. In the fourth choice, students could decompose the 8 into 4 and 4; then add one 4 to the 6 to make a 10. It is also possible that students would use the near double $6 + 6 = 12 \rightarrow 12 + 2 = 14$.

Try This

Have students find all the combinations of 10 within the context of a problem. An example is given as Appendix B, which can also be downloaded as a blackline master on NCTM's More4U site (nctm.org/more4u).

Decomposing a Number Leading to a Ten

- Sam baked _____ cookies. He ate _____ of them. How many cookies does he have left?

A	(12, 2)	(12, 3)	(12, 5)
B	(15, 5)	(15, 7)	(15, 8)

In the first number choice, the answer is 10, which students can then apply to the second and third number choices. There are two common ways students might decompose to a 10 for the number choice (12, 3): First, students could decompose the second number, or subtrahend, 12 – 2 = 10, 10 – 1 = 9; second, they could decompose the first number, or minuend, into tens and ones, 10 – 3 = 7, 7 + 2 = 9.

Using the Relationship between Addition and Subtraction

In using the relationship between addition and subtraction, it may be easier for students to use what they know from addition to help them with subtraction rather than use what they know from subtraction to help with addition. To introduce students to using this relationship, you may want to try posing two related types of problems, an add-to and a take-from, with similar number choices over the course of two days. In the first problem, the add-to, students would gain experiences with particular number choices that they could then apply on the second day to the take-from problem.

- Jack was preparing for a snowball fight. He already had _____ snowballs, and then he made _____ more. How many snowballs did Jack have?

    (2, 4)   (6, 5)   (8, 7)   (12, 4)

- Jack had _____ snowballs for a snowball fight. He threw _____ of them. How many does he have left?

    (6, 2)   (6, 4)   (15, 7)   (15, 8)

The numbers in the second problem are the same numbers that were used in the first day's work. For the number choice (6, 4), there are two relationships: one from the day before and one from the previous number choice. Students do not always remember what they did the day before; therefore, it is helpful to provide some reminders. For instance, you could leave the work from the previous day's sharing session in view and say something like, "When solving your problems today, see if there's anything that we did yesterday that could help you with today's problem. If there is something, tell me why that helps you."

Creating equivalent but easier or known sums

- Ellie received _____ cards on Monday. On Tuesday, she received _____ more cards. How many cards did Ellie receive?

  (4, 4)   (4, 5)   (7, 7)   (7, 9)

The numbers are selected to make use of the children's knowledge of doubles. Eventually, you would remove the scaffold of the previous number choice and pose different types of number choices to prompt students to use various strategies in a single problem. In other words, you want students to become independent in selecting strategies for different number combinations. Therefore, you want number choices that vary within a single problem, like those below:

- Ellie received _____ cards on Monday. On Tuesday, she received _____ more cards. How many cards did Ellie receive?

  (8, 4)   (7, 6)   (9, 7)

- Sam baked _____ cookies. He ate _____ of them. How many cookies does he have left?

  (13, 5)   (15, 9)   (17, 8)

1.NBT.4: Add within 100, including adding a two-digit number and a one-digit number, and adding a two-digit number and a multiple of 10, using concrete models or drawings and strategies based on place value, properties of operations, and/or the relationship between addition and subtraction; relate the strategy to a written method and explain the reasoning used. Understand that in adding two-digit numbers, one adds tens and tens, ones and ones; and sometimes it is necessary to compose a ten.

When addressing this standard, there are several different types of number combinations that you would want to expose your students to so that they notice and make use of number structures in their strategies. To illustrate the different types of number combinations, we provide a number progression that is divided into four categories (A–D). A number progression is a series of numbers that increase in complexity, though not necessarily in difficulty (Land and Drake 2014).

A	B	C	D
(10, 6) (30, 8)	(42, 8) (55, 9)	(40, 10) (60, 20)	(60, 27) (36, 40)

In the first category, the numbers involve adding a single-digit number to a multiple of 10. You want students to notice that when adding these two types of numbers that the sum consists of the digit in the tens and the digits in the ones places.

In category B, the numbers are a little more difficult because the two-digit number is not a multiple of 10. In addition, these number choices involve regrouping, which further increases the level of difficulty. Students could use strategies from the 1.OA.6 standard, such as make a ten, to help solve larger-number choices included in this problem. In the third category, the numbers involve two multiples of ten. Students should notice that the tens' place is changing and the ones' place is not. In the final category, students would be adding two two-digit numbers where one number is a multiple of ten and the other is not.

When starting to use a number progression, present several number choices from the first category and one from the second as we do in the problem below. This type of number choice sequence prompts students to notice the difference in the final number choice and to make connections to number structure.

- Ellie received _____ cards on Monday. On Tuesday, she received _____ more cards. How many cards did Ellie receive?

    (20, 4)   (40, 6) (90, 7)   (42, 6)

In our last number choice, we only changed the digit in the ones place from the multiple of ten in the second choice to set up students to make a connection that would scaffold their transition to a more difficult number choice.

1.NBT.6: Subtract multiples of 10 in the range 10–90 from multiples of 10 in the range 10–90 (positive or zero differences), using concrete models or drawings and strategies based on place value, properties of operations, and/or the relationship between addition and subtraction; relate the strategy to a written method and explain the reasoning used.

This standard addresses subtraction of multiples of ten only, but includes many strategies. Students must relate those strategies to a written method and be able to explain their reasoning. Here, the focus is on direct modeling and the relationship between addition and subtraction. Several problem types fit this standard.

Take-From Problem

- Jack had _____ snowballs for a snowball fight. He threw _____ of them. How many does he have left?

    (30, 20) (50, 10) (50, 30) (70, 30) (90, 40)

If students are direct modeling, which is very likely in first grade, they still might not be using tens consistently. In other words, students might use single cubes and count them by ones. Or, students might model the first number with tens, but take away the second number by ones.

> ## Try This
>
> Give students a combination of cubes grouped by tens and some single cubes. Observing which cubes students use initially tells you something about their understanding of "ten" as a unit.

Add To-Change Unknown Problem

- There are _____ fruit bats at the zoo. How many more fruit bats need to come to the zoo to have _____ bats?

    (1, 3)   (10, 30)   (10, 50)   (30, 70)   (40, 90)   (100, 300)

Students can easily use the relationship between addition and subtraction to solve this problem type. One of the equations matching this story is $10 +$ _____ $= 30$. Many times students prefer to use addition in this situation even though subtraction makes sense to some adults. As the teacher, however, you will want students to make explicit connections between addition and subtraction and to come to understand the relationship between the two operations.

Compare-Difference Unknown Problem

- Jack had _____ snowballs. Lily had _____ snowballs. How many more did Jack have?

    (20, 10)   (40, 20)   (50, 10)   (70, 60)

This problem type also allows students to use either addition or subtraction when solving, but it is more challenging so you should be prepared for some children to struggle. Because of the difficulty, start with the easiest possible number choices involving multiples of ten and make subsequent number choices more demanding by making the difference in the numbers greater. In the last number choice, the numbers are larger, but with a small difference, providing a scaffold into the larger numbers.

> ## Resources
>
> For more problems that meet the Number and Operations in Base Ten standards, see Appendix C. Problems in this appendix can also be downloaded as blackline masters at NCTM's More4U site (nctm.org/more4U).

2.NBT.5: Fluently add and subtract within 100 using strategies based on place value, properties of operations, and/or the relationship between addition and subtraction.

This standard is really broad because it focuses on all problem types and all numbers within 100. Therefore, it becomes the responsibility of the teacher to be purposeful in using number choices to prompt students to use more sophisticated strategies as well as to become flexible with their strategies. In Appendix C, we have provided an increasingly complex sequence of problem types and numbers that could help you develop a progression of problems and number choices that are appropriate for your students.

In second grade, more students start to use invented algorithms, or strategies, based on place value. You need to allow students to be creative, but you also want students to be accurate, to progress to more efficient strategies, and to be flexible when determining the strategy best suited for a particular situation, rather than using one strategy for every problem and number choice. Below are several problems with number choice progressions. You should spend a few days on similar number choices to provide students with opportunities to recognize and use patterns and relationships in their solution strategies. For instance, if students spend some time adding multiples of ten and understand what that means and what that looks like, they can apply that knowledge to adding other two-digit numbers.

Join To–Result Unknown Problem

- Last month, Luis collected _____ pennies. This month, he collected _____ pennies. How many pennies does Luis have now?

A	B	C	D
(30, 20)	(20, 35)	(36, 22)	(36, 38)
(20, 40)	(32, 40)	(27, 71)	(45, 37)
(10, 70)	(50, 16)	(43, 55)	(28, 59)

In the first column, only multiples of ten are provided as number choices. With enough encounters confined to multiples of ten, students will be able to use that knowledge with a more difficult number choice. In the second, one number in each choice is a multiple of ten while the other is not. The third column contains no multiples of ten; the fourth column contains no multiples of ten either, but when added together, the ones places do go over a ten.

Strategy A	Strategy B	Strategy C
counting on 36 + 22	36 + 10 = 46 46 + 10 = 56 56 + 2 = 58	30 + 10 = 40 40 + 10 = 50 50 + 6 = 56 56 + 2 = 58
Strategy D 30 + 20 = 50 6 + 2 = 8 50 + 8 = 58	Strategy E 36 + 20 = 56 56 + 2 = 58	

Fig. 5.2. Some strategies students might use to solve 36 + 22

We included strategy A to recognize that some students will need to count, but you want them to work toward more efficient strategies. The remaining strategies are based on place value or use of tens and ones. There are several unique ways that the numbers can be decomposed and put back together, and previous experiences with the various number types can help students use the diverse strategies.

Next, move away from join to–result unknown problems and provide take from–result unknown problems with number progressions.

Take From–Result Unknown Problem

- There are _____ Rockhopper penguins on the cliff. _____ of them jump off the cliff into the deep water. How many Rockhoppers are left on the cliff?

  (50, 30)   (55, 30)   (55, 33)   (50, 35)   (55, 38)

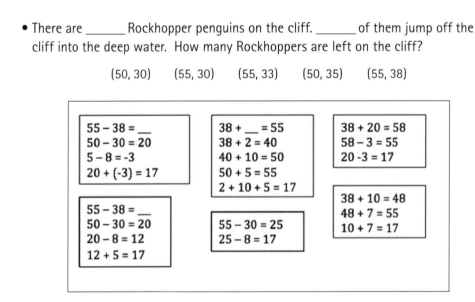

Fig. 5.3. Some strategies that students might use to solve 55 – 38

2.NBT.6: Add up to four two-digit numbers using strategies based on place value and properties of operations.

This standard is similar to 2.NBT.5, but it extends addition from two two-digit addends to up to four two-digit numbers. In these questions, you want students to employ the strategies that they have been using when adding two two-digit numbers and apply them to three and four addends. In the following problem, we chose numbers that would allow students to use a variety of strategies.

- The Rockhopper Penguin needs materials for its nest. It gathered _____ rocks, _____ pebbles, and _____ twigs. How many items did the Rockhopper gather?

(13, 21, 17)   (53, 24, 26)   (25, 25, 25,)   (35, 46, 45)

Students could use this strategy based on place value:

13 + 21 + 17 = _____
10 + 20 + 10 = 40
3 + 1 + 7 = 11
40 + 11 = 51

Or, they could use a strategy utilizing both place value and the commutative property:

13 + 17 + 21 = _____
10 + 10 + 10 = 30
30 + 21 = _____
30 + 20 + 1 = 51

3.OA.5: Apply properties of operations as strategies to multiply and divide. Examples: If 6 × 4 = 24 is known, then 4 × 6 = 24 is also known. (Commutative property of multiplication.) 3 × 5 × 2 can be found by 3 × 5 = 15, then 15 × 2 = 30, or by 5 × 2 = 10, then 3 × 10 = 30. (Associative property of multiplication.) Knowing that 8 × 5 = 40 and 8 × 2 = 16, one can find 8 × 7 as 8 × (5 + 2) = (8 × 5) + (8 × 2) = 40 + 16 = 56. (Distributive property.)

This is another standard that focuses on applying properties of operations. With multiplication, students can use the commutative, associative, and distributive properties. Just as in addition, you can choose numbers that make it more efficient to use these properties when solving problems similar to our examples below:

- There are _____ boxes of markers. Each box has _____ markers in it. How many markers are there?

(3, 12)   (12, 3)   (15, 5)   (21, 6)   (32, 7)

In the first two number choices, the numbers are simply reversed. This is to prompt students to think about reversing the numbers when solving. For most students, three groups of 12 is an easier problem to solve than 12 groups of 3. The situation is the same with the latter three number choices, in which it is easier to solve the problem by reversing the numbers. Once the numbers are reversed (solve for 6 groups of 21 rather than 21 groups of 6), the distributive property comes in handy by finding 6 groups of 20, then adding the 6 ones.

In the next problem, there are three factors that must be multiplied:

- There are _____ crates. Each crate has _____ bags. There are _____ oranges in each bag. How many oranges are in the crate?

$$(6, 3, 5) \quad (5, 3, 10) \quad (7, 2, 10) \quad (10, 9, 2)$$

After students solve this problem or one similar to it, you would want to have a conversation about the order in which students multiplied the numbers and if one order was easier than the others.

In the following problem, the number choices scaffold students' use of the distributive property; therefore, students must solve both number choices in the column selected.

- There are _____ boxes of markers. Each box has _____ markers in it. How many markers are there?

A	B	C
(3, 10)	(6, 20)	(4, 30)
(3, 11)	(6, 22)	(4, 33)

For each column, students can use what they know from working with multiples of ten to solve the first number choice, and then apply that to the second number choice. In the first column, a 1 was added to the 10 in hopes that students would notice that it is just "one more." After several encounters with number choices like these, you can remove the scaffold. Another, but more difficult, method for using the distributive property can be used with number choices in which the number in each group is reduced slightly in the second number choice from that of the first number choice as in the following problem:

- There are _____ boxes of markers. Each box has _____ markers in it. How many markers are there?

A	B	C
(3, 20)	(6, 30)	(4, 40)
(3, 19)	(6, 28)	(4, 37)

In each of these columns, the solution for the second number combination in the column can be found by using the first number combination and subtraction. For instance, in column A, the answer to (3, 19) is 3 less than (3, 20) because in (3, 19), there is one less in each group, and there are 3 groups.

True/false and open-number sentences (Carpenter, Franke, and Levi 2003) work well in helping students gain knowledge of and experience using the properties. For instance, you could pose something like the following true/false sentences to support understanding of the commutative property:

$$6 \times 4 = 24 \qquad \text{True or false? Why?}$$
$$4 \times 6 = 24 \qquad \text{True or false? Why?} \_\_\_\_\_$$
$$6 \times 4 = 4 \times 6 \qquad \text{True or false? Why? } \_\_\_\_\_$$

The following open-number sentences can help students gain understanding of the distributive property:

$$6 \times 4 = (3 \times 4) + (3 \_\_\_\_\_)$$
$$(10 \times \_\_\_\_\_) + (5 \times \_\_\_\_\_) = 15 \times 8$$
$$10 \times \_\_\_\_\_ = (10 \times 10) + (10 \times 3)$$

3.OA.7: Fluently multiply and divide within 100, using strategies such as the relationship between multiplication and division (e.g., knowing that 8 x 5 = 40, one knows 40 ÷ 5 = 8) or properties of operations. By the end of Grade 3, know from memory all products of two one-digit numbers.

While the previously discussed standard, 3.OA.5, focused on applying properties of multiplication, here the relationship between multiplication and division is highlighted. Many times, students find it easier to use multiplication to divide. Thus, for the problem below, a student might know that the answer is 8 for the first number choice because $3 \times 8 = 24$.

- There are _____ sugar cookies on the table. _____ cookies can fit on a plate. How many plates are needed for all of the cookies?

  (24, 3)   (50, 5)   (75, 5)   (100, 2)

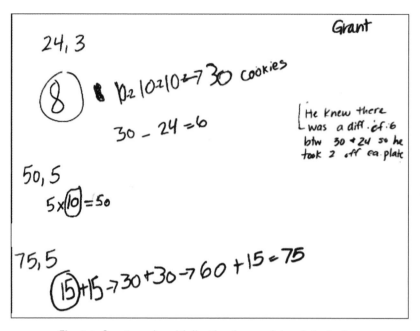

Fig. 5.4. Grant used multiplication in a variety of strategies.

When examining Grant's work, you see a variety of strategies that make use of multiplication. For (50, 5), 5 × 10 is a known fact for Grant. In the other two number choices, Grant makes use of repeated addition. He added three 10s for (24, 3), which got him to 30; then subtracted the 24 with a result of 6. Grant took care of the extra 6 cookies by taking 2 off each plate. In the last number choice, Grant is again adding repeatedly: two 15s is 30, another 30 would be four 15s, and then the fifth 15 is 75. It is unclear how Grant knew 15 in the first place. It might have been his first guess, or it might be that he did a similar adding process in his head with other numbers before the 15.

3.NBT.2: Fluently add and subtract within 1000 using strategies and algorithms based on place value, properties of operations, and/or the relationship between addition and subtraction.

To meet this standard, students must use strategies and algorithms, which is a slight change in the Common Core language from that of the previous standards discussed. Do not confuse the term "algorithm" with "standard algorithm." Throughout this book, our interpretation of the definition of *algorithm* is a method of solving problems that is based on place value, properties, and relationships. Both invented and standard algorithms meet this definition.

We offer a take from–change unknown problem with a variety of number choices. These choices provide students with opportunities to analyze the numbers and choose a strategy that works well with particular numbers, for example, compensating for (164, 99) and breaking apart by place for (331, 216).

- There were _____ books in the library. Some fantasy books were checked out. There are _____ books left. How many fantasy books were checked out?

| (81, 31) | (271, 177) | (287, 77) | (164, 99) |
| (331, 216) | (434, 136) | (525, 126) | |

Subtraction by place value	Compensating	Incrementing
81 – 31 = 50 I know 80 – 30 = 50	(287, 77) 77 → 87 87 + 200 = 287 200 + 10 = 210	(164, 99) 99 + 1 = 100 100 + 64 = 164 1 + 64 = 65
271 – 177 200 – 100 = 100 70 – 70 = 0 1 – 7 = -6 100 + 0 + (-6) = 94	(434, 136) 136 → 134 134 + 300 = 434 300 – 2 = 298	
331 – 216 300 – 200 = 100 30 – 10 = 20 1 – 6 = -5 200 + 20 + (-5) = 115	(525, 126) 525 + 1 = 526 526 – 126 = 400 400 – 1 = 399	
331 – 216 331 – 200 = 131 131 – 10 = 121 121 – 6 = 115		

Fig. 5.5. Strategies that students might use to solve the above problem

3.NBT.3: Multiply one-digit whole numbers by multiples of 10 in the range 10–90 (e.g., 9 × 80, 5 × 60) using strategies based on place value and properties of operations.

What is interesting about this standard is that it focuses on multiples of tens, which allows students to gain explicit experiences working with place value. The problem below provides an assortment of number choices that fit within this standard.

- Mrs. Franke has _____ boxes of pencils. Each box has _____ pencils in it. How many pencils does Mrs. Franke have?

      (8, 10)   (20, 5)   (40, 6)   (30, 7)   (10, 38)   (38, 10)

The difficulty is escalated across the number choices by increasing the number sizes, switching the order of the numbers, and including a non-multiple of ten (e.g., 38) to be multiplied by 10, which goes beyond the standard.

Commutative Property	Repeated Addition	Associative Property	Distributive Property
(20, 5) $5 \times 20 = 100$	(8, 10) $10 + 10 + 10 + 10 + 10 + 10 + 10 + 10 = 80$	(40, 6) $4 \times 6 \times 10 = 240$ Use of tens $4 \times 6 = 24$ $40 \times 6 = 240$	(38, 10) $30 \times 10 = 300$ $8 \times 10 = 80$ $300 + 80 = 380$

Fig. 5.6 Strategies students might use with the number choices for the problem above.

Students might use these strategies based on properties without formally stating the property in mathematical terms. At some point, a teacher would want to formally name these properties for students.

Conclusion

The Common Core is explicit about which strategies students should be using at certain grade levels, though it also lists multiple strategies in many standards. In addressing the Common Core, it is important to be mindful that each standard is an end-of-grade-level standard and that students must transition through a progression of strategies as they become more efficient with each one. For example, when first working with multiplication, students will need to direct model. As students continue working with multiplication, they will move on to counting strategies and invented algorithms (Carpenter et al. 1999). While your students progress through these various strategies, think about which numbers will either reinforce their use of a particular strategy or push them on to one that is more efficient.

Chapter 6

Relational Thinking and Number Choice

In recent years, there has been a shift in how mathematics educators think about algebra. "Algebra is no longer thought of as a subject but as a way of thinking and acting on mathematics objects, structures, and situations" (Molina, Castro, and Ambrose 2005, p. 265). Increased attention to algebra has resulted in part from the notion that it is the gatekeeper to higher mathematics (Moses and Cobb 2001). Much work has been done to make algebra more accessible to all students, such as including an algebra strand in the Common Core Standards from kindergarten through grade 12. In this chapter, we will demonstrate how to use number choice to promote relational thinking. "Relational thinking represents a fundamental shift from an arithmetic focus (calculating answers) to an algebraic focus (examining relations)" (Jacobs et al. 2007, p. 261). For example, consider the following number sentences:

$$2 \times 18 = 4 \times \underline{\hspace{1cm}}$$

$$100 + \underline{\hspace{1cm}} = 99 + 44$$

In both of these examples, students could calculate the answer ($2 \times 18 = 36$; $36 \div 4 = 9$), but they could also use the relationships in the number sentences to find the answer. For example, 4 is twice as much as 2, so the answer must be half of 18. Or 100 is one more than 99, so the answer must be one less than 44. When engaging in relational thinking, students are thinking algebraically and using the relationship between the numbers as a solution path.

In chapter 3 (page 31), we saw how Jack solved the Pennies problem: He changed the 97 to 100 as his first step. In other words, he saw that 97 was only 3 away from 100 and used that understanding to solve the problem. Students who use relational thinking are demonstrating different kinds of understandings about numbers than those of students using standard or invented algorithms. The chart on the following page outlines what can be deciphered about students' abilities on the basis of their use of the standard algorithm, an invented algorithm, or relational thinking to solve problems.

Standard Algorithm Students—	Invented Algorithm Students—	Relational Thinking Students—
• can get the right answer; • can add facts to 10; and • can carry out a procedure.	• can get the right answer; • understand place value; • can compose and decompose numbers; • can add basic facts to 10; and • can carry out a procedure.	• can get the right answer; • know basic facts and can use them to find others; • understand place value; • have a sense of equality; • understand conservation of large numbers; • are flexible in their thinking; and • can mentally compute answers.

Fig. 6.1. Students' uses of certain strategies indicate different mathematical abilities.

To be clear, we want to state that if a student uses the standard algorithm, it does not mean that he or she does not know and understand the concepts in the other two columns; it is only that the standard algorithm does not reveal as much about what students know.

In *Thinking Mathematically: Integrating Arithmetic and Algebra in Elementary School* (Carpenter, Franke, and Levi 2003), the authors provide information and strategies (the majority in the form of true/false and open-number sentences) for prompting students to think relationally. We recommend *Thinking Mathematically* as a resource that complements our work on number choice in promoting the development of students' relational thinking. In the following sections, we provide number choices to prompt relational thinking as well as true/false and open-number sentences to use with those number choices. We start with a first-grade standard because you want students to have a bank of known facts they can work with before you present content that introduces relational thinking. To progress from one level to the next, students must be able to relate something they do not know to something that they do. The basic facts are the foundation on which students build.

1.OA.1: Use addition and subtraction within 20 to solve word problems involving situations of adding to, taking from, putting together, taking apart, and comparing, with unknowns in all positions (e.g., by using objects, drawings, and equations with a symbol for the unknown number to represent the problem).

Adding One More

Within this standard, we illustrate how you can set up situations with number choice to prompt students to use relational thinking when solving problems. We started with the "adding one more" relationship, for example, 4 is one more than 3, because it is an easier (if not the easiest) relationship for students to recognize and use:

Add To–Result Unknown Problem

- Ellie received _____ cards on Monday. On Tuesday, she received _____ more cards. How many cards did Ellie receive?

A	B	C	D	E
(5, 5)	(7, 3)	(6, 4)	(18, 2)	(16, 4)
(5, 6)	(7, 4)	(6, 5)	(18, 3)	(16, 5)

You want the first number choice to be a known fact so that students have a solid understanding before they try to extend that understanding. Here, we are assuming that the first number combination in each column is a known fact for students. We want them to use the known fact to help themselves solve for the sum of the second combination. As you can see, the relationship between the first set of numbers in a column and the second set in that column is "one more."

Teacher Talk

In each of our problems, we included a "helper" number choice. Eventually, you want to eliminate the helper problem to find out if students use the relationships among the numbers without prompting or scaffolding.

A series of true/false and open-number sentences that might be posed to prompt relational thinking is on the following page. But before presenting the example, we offer some guidance on how and when to pose number sentences while first engaging students with relational thinking. It is helpful to involve students in relational thinking before posing a problem because we want the prior work and the attendant discussion to support the students as they work independently on a problem.

We usually pose the number sentences on a whiteboard or overhead one at a time to allow discussion of each sentence (see sample discussion below). This method allows the teacher to address any immediate needs and misconceptions that a particular sentence may raise. However, there are multiple ways to implement a true/false or open-number sentence task: Students could have personal whiteboards or notebooks in which they record their thinking or they could work with partners. If students have previously been engaged with relational thinking on multiple occasions, another option is to print out the entire series of number sentences for students to solve, and then follow up with a discussion. If students are familiar with relational thinking and with the mathematical concept being explored, you could pose all the number sentences at once and consider it an opportunity to practice. By varying your approach to this routine, students will continue to be engaged.

In the example problem below, we want students to think of $7 + 4$ as $7 + 3 + 1$. To reach this end goal, we would pose the following true/false and open number sentences:

Number Sentence	Rationale
$4 = 3 + 1$ True or false?	In this first sentence, we are asking students to consider decomposing 4 into 3 and 1.
$7 + 3 =$ _____	Here, we are highlighting two numbers that make a ten.
$10 +$ _____ $= 11$	We want students to recognize from the previous number sentence that 11 is one more than 10.
$7 + 4 = 7 + 3$ True or false?	Here, we want students to recognize that the equation cannot be true, because you cannot add two different numbers to 7 and get the same answer, but we also want them to recognize that 4 is one more than 3.
$7 + 4 = 7 + 3 +$ _____	Finally, we ask students to articulate that $7 + 4$ is the same as $7 + 3 + 1$ (or $6 + 4 + 1$; see Teacher Talk on the opposite page).

Fig. 6.2. The steps to prompting students to understand $7 + 4 = 7 + 3 + 1$

Below, we provide an example of a likely conversation in Jenny's classroom to illustrate the kind of teacher–student discussion using true/false and open-number sentences that supports students' work.

Mrs. Johnson: Look at this number sentence, 4 = 3 + 1, and try to decide if it is true or false. Remember, I am going to ask you to tell me why it's true or false and to give me a reason why you think that.

Sally: I think it's true because I know 3 + 1 = 4.

Mrs. Johnson: So because 3 + 1 = 4, you know 4 is the same as 3 + 1. Does anyone else agree with this?

Henry: That's true because it doesn't matter if the equal sign comes first.

Mrs. Johnson: Now look at this number sentence and tell me what you would need to put on the blank to make it true: 7 + 3 = _____.

Tad: I think it's 10 because I counted up on my fingers.

Mrs. Johnson:	How many did you count on your fingers? Can you say it out loud so I can hear you?
Tad:	7... 8, 9, 10
Cara:	I just knew 7 + 3 was 10.
Mrs. Johnson:	Lets look at the next number sentence, 10 + _____ = 11. What number do you think goes on the blank?
Manny:	I think it's 1 because 10 + 1 = 11.
Stella:	I think it's 1, too, because you just have to count one more.
Mrs. Johnson:	Now I'm going to write a number sentence and I want you to talk with your partner and decide if you think this number sentence is true or false: 7 + 4 = 7 + 3 How would you convince me (Carpenter, Franke, and Levi, 2003)?
Angel:	We think that it's false because 7 + 4 = 11 not 7.
Winston:	We thought it was false, too, because 7 + 4 is not the same as 7 + 3.
Mrs. Johnson:	How do you know 7 + 4 is not the same as 7 + 3?
Angel:	Well we know that 7 + 3 is 10 from up there, so 7 + 4 can't be 10.
Mrs. Johnson:	What is 7 + 4?
Angel:	It's 1 because it's just one more.
Mrs. Johnson:	Winston, what do you think of Angel's answer?
Winston:	I was thinking since the 7 came after the equal sign, that's why it was wrong.
Mrs. Johnson:	Now look at the last number sentence, 7 + 4 = 7 + 3 + _____. What do you think should go on the blank?
Annie:	I think it's 1 because 7 + 3 is 10 and 1 more would make 11.
Joey:	I think it's 1, too, but I know 3 + 1 = 4 and the 7s are the same so I didn't have to add those.
Mrs. Johnson:	What do you mean, "the 7s are the same"?
Joey:	Well, since there is a 7 on both sides, I don't need to add those. I can just look at the other numbers.

Teacher Talk

While I (Jenny Johnson) usually plan out five to six equations or number sentences to pose to students every day, sometimes our discussion leads me to present an unplanned equation to make either a connection or clarification for my students. Because of Joey's response, I was thinking an equation such as 32 + 10 = 32 + 7 + _____ would be good to pose.

Double + 1

Making and using doubles is another strategy commonly used in first grade because students generally know their doubles facts.

Put Together–Addend Unknown Problem

- There were _____ balls in the playground bucket. _____ of them were red. The rest were blue. How many blue balls were there?

A	B	C	D
(8, 4)	(12, 6)	(30, 15)	(36, 18)
(9, 4)	(12, 7)	(30, 16)	(36, 19)

Again, we would pose a series of true/false and open-number sentences to encourage students to make connections among the numbers:

Number Sentence	Rationale
1 + 6 = 6 + 1 True or false?	We want students to articulate that one can switch the numbers in addition and get the same answer.
_____ = 6 + 1	Here, we are focusing on the +1 part of doubles + 1.
6 + _____ = 12	We want students to use doubles to find the missing addend.
6 + 7 = 6 + 6 True or false?	Again, we want students to notice that you cannot add two different numbers and get the same answer. We also want them to articulate that 6 + 7 is one more than 6 + 6 because 7 is one more than 6.
6 + 7 = 6 + 6 + _____	Last, we ask students to articulate that 6 + 7 is the same as 6 + 6 + 1.

Fig. 6.3. The steps to prompting students to understand 6 + 7 = 6 + 6 +1

Transforming the Task with Number Choice

Another problem type that is appropriate to use with double and doubles +1 is presented in the following example:

Add To–Change Unknown Problem

- There are _____ fruit bats at the zoo. How many more fruit bats need to come to the zoo to have _____ bats?

A	B	C
(4, 8)	(6, 12)	(7, 14)
(4, 9)	(6, 13)	(7, 15)

Subtraction

Using relational thinking in a subtraction situation can be a bit more difficult for students. For instance, one student we worked with generated equations that equaled 25 and wrote the following:

- $100 - 75 = 25$

- $101 - 74 = 25$

- $102 - 73 = 25$

- $103 - 72 = 25$

In this series, the child tried to use relational thinking to generate her equations. Her misstep, however, was that when you increase the minuend (100) by one, you also need to increase the subtrahend (75) by one. This relationship is not intuitive in subtraction. Instead, she decreased the subtrahend by one.

Teacher Talk

In the Cookies problem on the next page, students might not know right away if the answer is one more or one less. The answer is one more because the minuend (first number) increased by 1 and the subtrahend (second number) stayed the same.

Consider the following problem that uses numbers that focus on subtracting from 10 to help students use that knowledge to subtract from other numbers:

73

Take From–Result Unknown Problem

• Sam baked _____ cookies. He ate _____ of them. How many cookies does he have left?

A	B	C
(10, 3)	(10, 4)	(10, 5)
(11, 3)	(11, 4)	(11, 5)

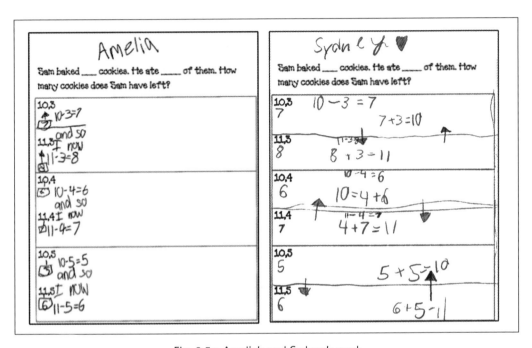

Fig. 6.5a. Amelia's and Sydney's work

If a student knows 10 – 3 = 7, then they can use that information to help them solve for 11 – 3. Here and on the following page, we share the work of three first-grade students: Amelia, Sydney, and Will. In both Amelia and Sydney's work, we see evidence that they used the first fact in each column to help with the second. Amelia, new to relational thinking, wrote, "and so I now [know]." Sydney indicated that she noticed a relationship with arrows. Also interesting in Sydney's work is her inclusion of equations that denote the inverse operation. In the last two number choices, Sydney uses addition only. Will's thinking is a little different in that he works within and across the number choices as denoted by his arrows. He used the first column of number choices to help with the second. For the last column of number choices ((10, 5) and (11, 5)), Will writes 10 + 1 = 11 vertically, using the 10 from 10 – 5 = 5, and then uses the 11 to start 11 – 5 = 6. This is further evidence that Will saw a relationship between the two sets of numbers.

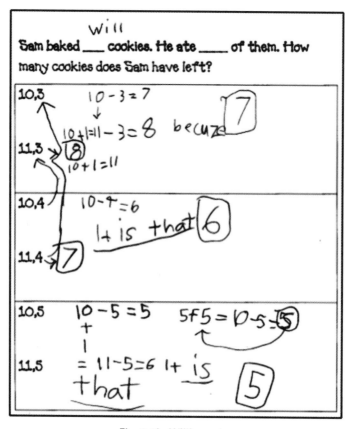

Fig. 6.5b. Will's work

In the next problem, we are employing 10, but using it when subtracting from 9:

• Sam baked _____ cookies. He ate _____ of them. How many cookies does he have left?

	A	B	C
	(10, 3)	(10, 4)	(10, 5)
	(9, 3)	(9, 4)	(9, 5)

In these cases, the answer to the second number choice in each column is going to be one less because you start with one less.

Using the relationship between 10 and numbers between 11 and 19 is another useful way to solve problems. Consider the same question with the following number choices:

• Sam baked _____ cookies. He ate _____ of them. How many cookies does he have left?

	A	B	C
	(14, 10)	(16, 10)	(18, 10)
	(14, 9)	(16, 9)	(18, 9)

The hope is that because of prior work with K.NBT.A.1 (compose and decompose numbers from 11 to 19 into ten ones and some further ones), students will be able to easily subtract 10 from a teen number. Having this knowledge, students could also subtract 9 if they correctly recognize that subtracting one less results in an answer of one more.

The next problem focuses on making use of doubles when subtracting:

• Sam baked _____ cookies. He ate _____ of them. How many cookies does he have left?

	A	B	C
	(12, 6)	(12, 6)	(14, 7)
	(13, 6)	(12, 7)	(14, 8)

In both columns A and B, we make use of double 6, but in two different ways: In A, we increase the 12 by 1, and in B, we increase the 6 by 1. By doing this, we are encouraging students to be flexible with the relationship.

2.OA.2: Use addition and subtraction within 100 to solve one- and two-step word problems involving situations of adding to, taking from, putting together, taking apart, and comparing, with unknowns in all positions, e.g., by using drawings and equations with a symbol for the unknown number to represent the problem.

Addition

For the problem below, we chose numbers for each row that all had a relationship to other numbers in the row. We labeled the rows to indicate that a student should solve for all the numbers in a row.

• Ellie received _____ cards on Monday. On Tuesday, she received _____ more cards. How many cards did Ellie receive?

A	(6, 6)	(16, 6)			
B	(25, 6)	(25, 16)	(45, 16)		
C	(39, 8)	(38, 8)	(39, 18)	(99, 18)	(48, 18)

In row A, we want students to see that the second number choice is just ten more than the first so they could just add 10 to their first answer. We repeated this idea in row B, but added 10 to the second number instead. The first number of the last number choice in Row B is 20 more than that of the second number choice. Row C is different: The numbers are higher; the first number of the second number choice is one less than that of the first, the third and fourth number choices relate to the first, and the last number choice is related to the second.

2.OA.2: Making tens across single-, double-, and triple-digit numbers using relationships.

For this standard, we want to demonstrate making ten in addition across multi-digit numbers. If students establish a solid foundation of adding tens, they can use that knowledge to solve problems with double- and triple-digit numbers. The two problems below illustrate increasingly difficult number choices that involve making a ten. In the first situation, we focus on numbers less than 100; in the second, numbers greater than 100. Using what you know about numbers less than 100 for numbers greater than 100 is an important transition for students. Students might not always see that connection, reverting back to other methods when they encounter larger numbers.

- Last month, Luis collected _____ pennies. This month, he collected _____ pennies. How many pennies does Luis have now?

A	(6, 4)	(16, 4)	(16, 34)
B	(7, 3)	(17, 3)	(13, 7)

- Last month, Luis collected _____ pennies. This month, he collected _____ pennies. How many pennies does Luis have now?

A	(106, 4)	(106, 14)	(196, 14)
B	(53, 47)	(153, 147)	(247, 753)

You can see that we kept the focus on 6 + 4 and 7 + 3 for both problems. If working with larger numbers is your goal, then using the numbers in the first problem as a scaffold would be appropriate.

A series of true/false and open-number sentences that could be posed to help students notice the make-a-ten strategy with double-digit numbers is on the next page. We focused on decomposing 8 into 5 and 3; however, if students needed an easier number to decompose, 5 could be the focus instead:

Number Sentence	Rationale
5 + _____ = 8	In this series, we start off with a simple equation that establishes 5 + 3 = 8.
8 + 5 = 8 + 2 + _____	We move into equation work that involves decomposing 5 into 2 and 3.
5 + 8 = 8 + 2 + _____	In this equation, we switch around the numbers to help students be flexible.
25 + 5 = _____	Here, we prompt adding the 25 and 8 by decomposing the 8 into 5 and 3 and making a ten with the 5.
30 + _____ = 33	We focus on what must be added to 30 to get 33 by adding the 3 from the decomposed 8.
25 + 8 = 25 + 5 + 3	Finally, we want students to articulate the decomposition of 8 when adding 25 and 8.

Fig. 6.6. The steps to prompting students to understand 25 + 8 = 25 + 5 + 3

Subtraction with One or Two More

In the problem below, we focus on students recognizing relationships between numbers that are one or two more.

- Sam baked _____ cookies. He ate _____ of them. How many cookies does he have left?

A	B	C
(40, 13)	(50, 25)	(80, 50)
(41, 13)	(51, 25)	(80, 48)

If students can think about decomposing 13 into 10 and 3 in the first number choice in column A, then students could think about 40 − 13 as 40 − 10 = 30, 30 − 3 = 27. For the second number choice in column A, students would need to recognize that they started with a number that is one more so the answer must be one more, or 28. The latter two columns depict similar relationships.

3.OA.3: Use multiplication and division within 100 to solve word problems in situations involving equal groups, arrays, and measurement quantities, e.g., by using drawings and equations with a symbol for the unknown number to represent the problem.

Like other standards we have encountered, the one above could be addressed with many different number choices. It is also one for which you would want to use a number choice progression. We pose fairly low numbers below, but we would gradually increase the number sizes over the course of third grade.

- Sam had _____ fishbowls. He had _____ fish in each bowl. How many fish did he have?

A	B	C	D
(2, 6)	(2, 8)	(3, 11)	(3, 15)
(4, 6)	(4, 8)	(6, 11)	(6, 15)
(8, 6)	(8, 8)	(7, 11)	(7, 15)

This Fishbowl problem involves equal groups as per the standard, and involves a doubling relationship in each column. In columns A and B, you want students to notice that the answer to 4×6 is a double of 2×6 and that the answer to 8×6 is a double of 4×6. You might pose questions such as "What did you notice about these numbers?" "What in the story is staying the same?" "What is changing? "How is it changing?" "How does that affect the answer?" If students notice the doubling relationship and can use it in solving problems, they can also use that knowledge to build fact fluency. Numbers in columns C and D were chosen for several reasons:

- To provide students with an opportunity to notice the relationship between 3 and 6

- To see what happens when you multiply single-digit numbers by 11

- To notice that 7 times a number is just one more group of that number than 6 times a number

The problem on the next page has numbers that involve the relationship between multiplying by 10 and multiplying by 9. In this situation, we kept the group size the same in each column, but changed the number of groups from 10 to 9. This structure lets students think about one less group, which is a little easier than thinking of one less in each of the groups.

- Sam had bags of fruit. Each bag had _____ pieces of fruit. If he had _____ bags, how many total pieces of fruit did he have?

A	B	C
(2, 10)	(6, 10)	(8, 10)
(2, 9)	(6, 9)	(8, 9)

Students usually come to know their "10s facts" fairly quickly because of experiences with money, skip-counting by tens, and the apparent patterns in each. Therefore, it is useful to build on that knowledge for facts of other multiples.

We collected student work for the problem above from ten students. Four students used relational thinking in their strategies, which is expected. Not all students will use relational thinking, as they may not be ready to do that kind of thinking.

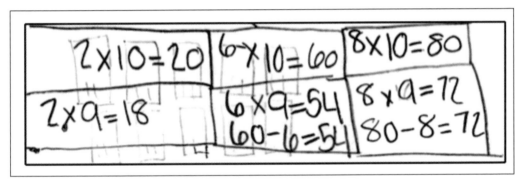

Fig. 6.7. Lesley's work

To prompt more students to use relational thinking, you may want to have students who did share their strategies. In the student work above, we see that Lesley easily multiplied the two factors together for the first column of number choices. She may not have used relational thinking for 2×9 because that was a known fact for her. For the other two columns, Lesley used what she knew about 6×10 and 8×10 to solve for 6×9 and 8×9. (Note: It is typical for Lesley to write an equation that matches the problem, but leave the answer blank until she reaches a solution.)

In Michelle's strategy on the following page we see her using relational thinking in a different way. Like Lesley, she wrote the equation first, and then sought the answer. Consider her work for 6×9. First, she switched the order of the factors. Next, she found $9 \times 3 = 27$, then added $27 + 27$, which indicated to us that she knew 9×3 was half of 9×6. Michelle used the same process for 8×9. Although Michelle did not use the relationship we prompted for, she found a different relationship among the numbers.

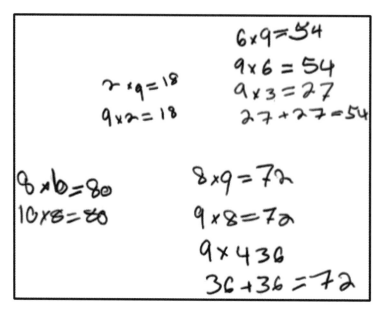

Fig. 6.8. Michelle's work

We use the Fishbowl problem again below. This time, we focus on multiplying by decade and hundreds numbers.

- Sam had _____ fishbowls. He had _____ fish in each bowl. How many fish did he have?

A	B	C
(3, 5)	(2, 8)	(6, 4)
(30, 5)	(2, 80)	(6, 40)
(3, 50)	(2, 800)	(60, 40)

Column A has numbers that require students to consider the relationship between 3 and 30 and 5 and 50—multiplication by 10. In other words, 30 is 3 × 10 and 50 is 5 × 10. We do not want students to simply "add a zero" at end of each number. Instead, we want them to notice the patterns when multiplying by multiples of ten and a hundred and to be able to articulate that "adding a zero" is really multiplying by 10.

Conclusion

In this chapter, we discussed relationships between numbers (e.g., +1, -1, doubling, etc.) that you can emphasize in your teaching of mathematics. There are, of course, many other relationships that you could highlight and students could use. The key is to pose problems and numbers that encourage students to think relationally and to use what they know to solve problems they do not know. In other words, once students have a solid base of known facts, present number combinations that have a relationship to the known facts and observe what students do. The most important and powerful relationships for students to learn and use involve properties and patterns of operations and of the base-10 number system, including the commutative property and patterns related to adding ten or multiplying by ten. Understanding and using these foundational relationships will provide students with access to higher-level mathematics, in algebra and beyond.

Chapter 7

Assessment and Number Choice

Number choice, when used in conjunction with assessment, opens up opportunities to focus not only on the Common Core standards but also to achieve them. It provides another approach to assessing any standard and verifying if the child meets that goal or not. By providing a progression of number choices that increases the difficulty of the problem type and the intricacy of solution strategies, the teacher can monitor the level of complexity the student is using to comprehend a standard. The interaction of these components can determine not only if the child has met the goal but also reveal important additional information on each child's understanding and where it falls on the level of progression. In this chapter, we discuss how we use number choice to assess students, and provide some tools to record students' strategies related to number choice.

Assessment and Addition

When starting a new unit, problem type, or standard, we suggest giving a problem with a single number choice that is carefully chosen to focus the assessment. This type of assessment is typically known as formative assessment. For example, consider the following standard:

2.NBT.5: Fluently add and subtract within 100 using strategies based on place value, properties of operations, and/or the relationship between addition and subtraction.

To initially assess students' understanding of the standard, we wrote the following add to–result unknown problem:

- Jack made 38 snowballs for a snowball fight and Tommy made 44 snowballs. How many total snowballs did they make?

The number choices 38 and 44 are good ones because they are within the standard, and they involve the ones place going over ten. This number selection fully meets the target standard while easier number choices like (6, 4), (20, 30), or (32, 50) do not. Numbers that fully meet the targeted standard provide more information about students' understandings.

Teacher Talk

If you have a student who cannot engage in this problem because the number choice is too difficult, it is important to immediately present an alternate choice to that child. For example, try (20, 10) or (12, 10). If a student demonstrates a high level of proficiency, posing a challenging number choice such as (87, 68) or (157, 76) will provide information on how they engage with larger numbers.

When giving this problem for preassessment purposes, our suggestion is to pose the problem to six to eight students, and take notes and ask questions to fully grasp their understanding. To help organize your notes, we suggest using a chart like the one below. We provide a page-length version of the table in Appendix D; it is also available to download as a blackline master on NCTM's More4U page.

Problem: Jack made 38 snowballs for a snowball fight, and Tommy made 44 snowballs. How many total snowballs did they have?

Key questions to check for understanding:
How did you decide to solve this problem?
What does _____mean to you?
What is the solution to this problem?
How did that sound as you were counting?

Direct Modeling	Counting	Invented/Standard Algorithm Correct	Invented/Standard Algorithm Incorrect
Student and details:	Student and details:	Student and details:	Student and details:

Fig. 7.1. Sample organizational chart

Depending on the grade level and problem, you might want to change some of the column headings to match the strategies for that particular problem. For the snowball problem, a teacher might see the following strategies being used by students:

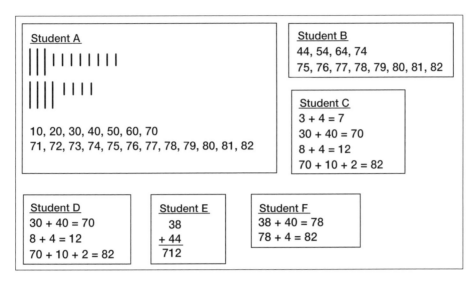

Fig 7.2. Student strategies for the Snowball I problem

Here is an example of where the work of the students above would fall in an organizational chart.

Problem: Jack made 38 snowballs for a snowball fight, and Tommy made 44 snowballs. How many total snowballs did they have?

Key questions to check for understanding:

How did you decide to solve this problem?

What does _____ mean to you?

What is the solution to this problem?

How did that sound as you were counting?

Direct Modeling	Counting	Invented/Standard Algorithm Correct	Invented/Standard Algorithm Incorrect
Student **A** models with tens and ones.	Student **B** uses a counting strategy.	Student **C** uses tens and ones. Student **D** also uses tens and ones. Student **F** uses incrementing.	Student **E** uses the standard algorithm.

Fig. 7.3. Student strategies for the Snowball I problem recorded in an organizational chart

Student	What We Learned	Numbers to Try Next	Questions/Other Support Strategies
A	• This student needs manipulatives to create the number and provide a tangible counting experience. • He is not meeting the standard.	Use single-digit numbers followed by adding that same number of tens (e.g., 4 + 3, and then 40 + 30) to prompt him to add tens together more efficiently rather than counting.	After posing 4 + 3, ask, "What if I had 4 tens and 3 tens; how many tens would that be?"
B	• This student is able to count by tens from a two-digit number other than a decade number. • She is not meeting the standard.	She may need to work with multiples of 10 to start working toward a more efficient strategy like breaking apart by place used by students C and D.	After working with multiples of 10, pose choices that involve one non-multiple of 10, for example, 44 + 20, 65 + 30, 20 + 48.
C	• This student is meeting the standard. • He uses what he knows to solve what he doesn't know, for example, 3 + 4 for 30 + 40. • He is not comfortable keeping the tens as tens.	For this student, you could go beyond the standard by posing numbers whose sum is greater than 100.	Watch for this misconception that because 8 + 4 = 12, then 80 + 40 = 112
D	• This student is more confident than C with addition of tens. • She decomposes numbers to make them easier to add.	As with C, you could pose numbers with sums greater than 100 or numbers that would encourage compensating, such as 39 + 23 or 49 + 35, or relational thinking.	
E	• Makes a common mistake with the standard algorithm, and indicates a lack of place-value understanding. • The student is unsure of what he or she knows.	Pose the same numbers as for A and B.	Ask this student to use cubes to show what he or she did.
F	• This student uses a fairly sophisticated incrementing strategy, showing a high level of flexibility and place-value understanding.	Pose the same numbers as for C and D.	

Fig. 7.4. Student assessment chart for Snowball I problem

Children can make errors when using any strategy. Common mistakes are counting errors or computation inaccuracies. For instance, it would be easy for students A or B to lose track of their counting, or for the other students to make a calculation error. In addition, a student employing direct modeling like student A might use tens inconsistently, for example, modeling the numbers with tens but then counting by ones.

There are several items to decipher about the students using these strategies; they are outlined in the chart on the preceding page. The students' level of sophistication of strategy usage and of attention to precision with counting and computation suggest future number choices. The goal of the teacher is to solidify students' strategies, and then move them to more efficient ones.

After analyzing the assessment, the following problems and number choices could be posed to the students to provide experiences that would correlate to the learning goal for each student.

Add To–Result Unknown Problem

- Jordan has _____ toy cars. He collected _____ more throughout the year. How many toy cars does he have now?

    (4, 3)   (40, 30)   (44, 30)   (67, 50)   (99, 57)   (109, 57)

Put Together/Take Apart–Total Unknown Problem

- _____ red tulips and _____ yellow tulips were growing in a garden. How many tulips were there in the garden?

    (4, 5)   (50, 40)   (56, 40)   (56, 44)   (96, 44)   (104, 66)

Add To–Change Unknown Problem

Sarah planted _____ tulip bulbs on Monday. On Tuesday she planted some more. Now Sarah has _____ tulip bulbs planted in her garden. How many bulbs did Sarah plant on Tuesday?

    (5, 9)   (50, 90)   (40, 92)   (52, 96)   (56, 92)   (96, 152)

Assessment and Subtraction

Another problem that could be used as a formative assessment to check for under-standing of standard 2.NBT.5 is a take from–result unknown problem:

- Jack had 63 snowballs for a snowball fight. He threw 37 of them. How many does he have left?

The numbers were selected to fully meet the standard. They are non-decade numbers within 100 that would traditionally involve regrouping. Particularly in this problem, you would want to ask questions about how students took away 7 and not simply accept, "I took away seven," without probing further.

Teacher Talk

If you have a student who cannot engage in this problem because the number choice is too difficult, it is important to immediately present an alternate choice to that child. For example, try (40, 10) or (17, 10). In addition, if a student demonstrates a high level of proficiency, posing a challenging number choice such as (77, 28) or (206, 158) will provide information on how he or she engages with larger numbers that have complex relationships.

As in the Snowball I problem, some students (A and B) might use direct modeling with ones or with tens and ones to solve the Snowball II problem. A teacher might also see students use other strategies.

Student C	Student D	Student E	Student F	Student G
63	63	60	$60 - 30 = 30$	$37 + \underline{} = 63$
$- 37$	$- 30$	$- 30$	$3 - 7 = -4$	$37 + 3 = 40$
34	33	30	$30 + (-4) = 26$	$40 + 20 = 60$
	$- 7$	$- 7$		$60 + 3 = 63$
	26	23		$3 + 20 + 3 = 26$
		$+ 3$		
		26		

Fig. 7.5. Student strategies for Snowball II problem

Problem: Jack had 63 snowballs for a snowball fight. He threw 37 of them. How
many does he have left?

Key questions to check for understanding:
How did you decide to solve this problem?
What does _____ mean to you?
What is the solution to this problem?
How did that sound as you were counting?

Direct Modeling	Counting	Invented/Standard Algorithm Correct	Invented/Standard Algorithm Incorrect
Student A correctly direct models. Student B incorrectly direct models.		Student D is incrementing using tens and ones. Student E is compensating using tens and ones. Student F is using tens and ones as well as negative numbers. Student G solves the problem using the inverse operation with an incrementing strategy.	Student C uses a standard algorithm, but has some misconceptions.

Fig. 7.6. Student strategies for the Snowballs II problem recorded in an organizational chart

Student C's strategy depicts a common misconception of students: They think they can reverse the numbers in subtraction as they do in addition. In our experience, many students will use addition to subtract because they find it easier. The chart on the following page describes what is learned about each student, what numbers to try next, and what other questions and strategies could help advance students' thinking, strategy level, and attention to precision.

Student	What We Learn	Numbers to Try Next	Questions/Other Support Strategies
A	Student can take a larger amount of ones from a smaller amount by taking away from the ten sticks.	• 60, 30: Do they direct model or is this known? • 63, 30 • 60, 32	Ask, "How did you take away 7?"
B	Student may not see the difference in value between a ten and a one, or they had an incorrect count, or both.	60, 30 Direct model, using base-ten pieces and encourage counting back.	Number work that includes counting back, for instance, 84, 74, 64, and so on.
C	Student is trying to apply commutative property to subtraction.	Encourage the use of base-ten blocks to direct model with various number choices, for example, (52, 38) and (63, 30).	• Ask, "How would you use these blocks to find your answer?" Pose true/false number sentences like 3 – 7 = 7 – 3.
D	• This student is meeting the standard. • The student decomposes 7 to make subtraction easier.	Three-digit numbers minus two-digit numbers, for example, 120 – 48 and 131 – 57.	• If children are easily able to work with these numbers, encourage them to try a second strategy. • Ask students F or G to share their thinking.
E	This student can decompose 63 and remembers to add the 3 back on.	Same numbers as for student D.	Same as D.
F	This student can decompose by place value and can use negative numbers.	The same numbers for student D will work for student F as well because the result would be negative numbers in the ones and tens places.	Start working on flexibility. Pose 120 – 40 before 120 – 48 to see if he or she can use straight subtraction and not break apart by place.
G	This student is very flexible in his or her understanding of subtraction, and meets the standard more fully than the others as he or she uses the relationship between addition and subtraction.	• Pose numbers greater than 100. • Pose numbers that are further away from each other to see how he or she adds in chunks.	Build flexibility to see if the student's strategy changes as the numbers get further apart.

Fig. 7.7. Student assessment chart for Snowball II problem

After analyzing the above assessment, the following problems and number choices could be posed to the students to provide experiences that would correlate to the learning goal for each student.

Take From–Result Unknown Problem

- In the same snowball fight as Jack, Jose had _____ snowballs for a snowball fight. He threw _____ of them. How many does he have left?

    (60, 30)   (63, 30)   (72, 30)   (120, 40)   (120, 48)   (253, 72)

Compare–Difference Unknown Problem

- Katie made _____ snowballs. Jessica made _____ snowballs. How many more snowballs did Katie have than Jessica?

    (30, 10)   (62, 20)   (70, 45)   (76, 37)   (137, 76)

Put Together/Take Apart with Addend Unknown Problem

- _____ snowmen were on the playground. _____ melted and the rest did not melt. How many snowmen did not melt?

    (40, 30)   (61, 20)   (50, 26)   (43, 27)   (129, 42)

Assessing Daily Student Work

Posing problems on a daily basis allows teachers to better understand the journey children take in developing their mathematical thinking and relationships. Crafting multiple problems with numbers selected to reinforce and scaffold the students' levels of engagement with the standard becomes the role of the teacher. The goal is to focus on what each child knows and demonstrates on a daily basis as opposed to relying solely on external tests and assessments. As the teacher, your responsibility is to sort through the students' work and words to determine a series of problems and number choices that work in tandem to advance the understanding of every child in the classroom. This complex interaction is challenging, but the knowledge gained from careful, daily attention to the students' ability to solve problems not only promotes a higher level of mathematical reasoning but also develops a classroom of perseverance.

Sometimes, students will change their strategy based on the number choice. For instance, consider the problem adapted by Molly for a third-grade lesson from *Investigations* (TERC 2008) and student work below (Drake et al. forthcoming):

• Amber went to the sticker station and she bought _____ soccer and _____ dance stickers. How many stickers did she buy?

(2, 8)   (25, 10)   (37, 30)   (52, 60)   (110, 71)   (241, 189)

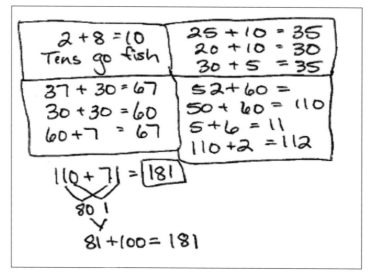

Fig. 7.8 "Tens go fish" in Olivia's solution for (2, 8) in the Sticker problem refers to a fact game played in the classroom.

Olivia solved for several of the number choices. (Olivia typically writes the equation that matches the story, leaves the answer blank, solves the problem, and then goes back and fills in the answer.) As you can see, Olivia changed her strategy twice based on the number choice. For (25, 10) and (37, 30), she broke apart the numbers by place. But (52, 60) was different because it went over 100. She broke apart by place, but then used 5 + 6 to help her with 50 + 60. Olivia changed her strategy again when solving for (110, 71). In that number choice, there is one three-digit number. Olivia still broke the numbers apart by place, but recorded that process with a "tree." All of these methods are great strategies to use and meet the Common Core Standards for third grade, but it is noteworthy that Olivia changes her strategy based on the number choice, indicating where she is comfortable.

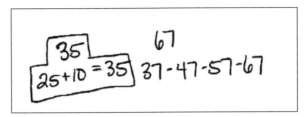

Fig. 7.9. Matt's work for the Sticker problem

Matt, a second-grader, solved the sticker problem for (25, 10) and (37, 30). For (25, 10), Matt could easily add 10 to 25 and used an equation to record his thinking. When confronted with adding a multiple of 10, however, he recorded his skip counting by tens. Matt is not quite comfortable enough to write 37 + 30 = 67 as a known fact.

In both of these cases, the student work provides important information about number choice. Olivia is quite comfortable with two-digit numbers that do not go over 100, but not as comfortable with numbers that do, which indicates that she should be working with numbers that go over 100. Matt most likely needs to solidify his understanding of adding multiples of 10. This type of information allows the teacher to match number selection in response to the students' work.

Conclusion

Every problem that a teacher poses to a student is the beginning of an assessment. Assessment should not be viewed as a finite exercise to determine mastery of skills sets or just the right answer but as a method to monitor the development of cognitive understanding in mathematics for every child in the classroom. Students will fall above or below the set grade-level expectations, so teachers need to be mindful that this range must be addressed when presenting children with problems for assessment. The process of aligning the standards with assessment and instruction can be accomplished by formulating a six-step plan:

1. Begin with a Common Core Standard that you want to address.

2. Select a problem type and write a quality problem that will meet the standard.

3. Choose a set of numbers that meets the standard.

4. Pose the problem to your students. Be a careful observer of their work, and ask questions.

5. Analyze their work, and recognize and identify the strategies that are being used within the classroom.

6. Create a bank of problems with number choices that will help each student progress forward toward his or her goal.

As you do this work on a consistent basis, you will see groups of children that use similar strategies spontaneously emerge in your classroom. For instance, you will have a group of children who direct model for addition, another group that uses invented algorithms, and so on. As you learn more about your students, you will be better able to refine your number choices to match the varied levels of your students. Try this six-step approach, and see how it transforms your assessment and instruction planning to deeply affect your students' mathematical thinking in immeasurable ways.

Final Words

Because mathematics teaching has changed very little over the past hundred years (Stigler and Hiebert 2009) and because of our personal observations of modern-day mathematics classrooms, we contend that our childhood experience of elementary mathematics is still the primary experience of many students in the United States. We remember answering two or three word problems at the end of several computation problems. Typically, we would complete twenty to thirty computation problems for one operation, for example, addition, and then solve the word problem by applying the same operation used in the computation problems. Little attention was given to number choice other than denoting single-, double-, or multidigit numbers along with numbers that needed to be carried or borrowed across zeros or not.

In 2010, Lampert and colleagues wrote that posing word problems is an instructional activity that is ubiquitous in elementary classrooms. The ways in which we have talked about number choice throughout this book can disrupt the very traditional use of word problems that is still prominent today. With a focus on number choice, we can adapt word problems to meet not only the content, strategies, and mathematical practices inherent in the Common Core but also to create mathematically rich classrooms by providing differentiation in a way so that all students are working productively, multiple learning goals are being met within one problem, and the mathematical needs of students are being addressed.

As you work with number choices, you will not only continue to learn about number choice but also be engaged in *generativity*. Generativity refers to—

> "the teachers' ability to continually add to their understanding by connecting their personal and professional knowledge with the knowledge they gain from their students in order to produce or originate new knowledge that is useful to them in pedagogical problem solving and in meeting the educational needs of their students" (Ball 2009, p. 47).

It is hoped that this book has added to your understanding of number choice, and we expect that you will soon gain additional knowledge of number choice through your own teaching practice with your students. Good luck!

Appendix A: Apple Problem with Ten-Frame

The farmer can fit 10 apples in a box. Some are red apples and some are yellow apples. What are the different ways the farmer can box the apples?

Appendix B: Finding All Combinations of 10

There are 10 bears in the bed. _____ fell out . How many bears are left in the bed?

10	0
3	2
7	6
1	4
8	9
5	

Appendix C: Number Choice Problems

Use the following problems with number choices to help your students develop understanding of and meet the stated Common Core Standard (NGA and CCSSO 2010). The set of problems for each standard should be posed to students within a short period of time (one to two weeks) to allow students to make connections. Consistent with the philosophy of this book, the number choices were crafted to give students multiple experiences to enable meaningful discussion of their strategies. To provide more interest in the problems, educators may want to exchange the names used in the problems with those of the students in their class.

1.OA: Represent and solve problems involving addition and subtraction.

- Horton sat on _____ eggs. _____ hatched. How many eggs did not hatch?

  (10, 3)   (42, 20)   (50, 3)   (48, 22)   (64, 48)

- My book has _____ pages. I have read _____ pages. How many more pages do I need to read?

  (10, 7)   (60, 20)   (64, 20)   (64, 22)   (100, 86)

- Mr. Brown made _____ sounds. Then he made _____ more sounds. How many sounds did Mr. Brown make?

  (3, 7)   (18, 20)   (40, 24)   (35, 41)   (23, 47)

- One animal had _____ feet in the front and _____ feet in the back. How many feet does it have?

  (13, 7)   (7, 80)   (80, 40)   (83, 44)   (79, 40)

- Sam _____ green eggs for breakfast, _____ green eggs for lunch, and _____ green eggs for supper. How many green eggs did Sam eat?

  (3, 5, 7)   (10, 7, 3)   (20, 4, 20)   (25, 5, 10)   (21, 32, 43)

- Gerald McGrew counted 14 feet at his zoo. What are some of the different animals he has at his zoo? List the animals' names and how many animals he has.

1.OA: Understand and apply properties of operations and the relationship be-
tween addition and subtraction.

- Jack saw _____ starfish. Each starfish had 5 arms. How many arms did Jack count in all?

  (2)   (4)   (5)   (10)

- Blake collects hermit crabs. He found _____ in the water and _____ on the beach. How many did he find in all?

  (8, 3)   (3, 8)   (9, 4)   (4, 19)

- Will saw _____ sand dollars. _____ of them broke. How many were not broken?

  (12, 8)   (12, 3)   (14, 7)   (24, 7)

- Talan saw _____ sea horses. Each sea horse had _____ babies in its pouch. How many babies did Talan see?

  (5, 2)   (2, 5)   (10, 2)   (2, 10)

- A pirate found some treasure. There were _____ red jewels, _____ blue jewels, and _____ green jewels. How many jewels did he find?

  (3, 8, 5)   (5, 8, 3)   (4, 5, 8)   (15, 18, 13)

1.NBT: Understand place value.

- There were _____ little clouds in the sky. It got cloudier. Now there are _____ little clouds in the sky. How many clouds were added?

    (10, 30)   (20, 48)   (12, 42)   (20, 120)

- The click beetle tried many times to flip over. First he tried _____ times. Then he tried _____ more times. How many times did the click beetle try to flip over?

    (10, 35)   (80, 24)   (24, 60)   (37, 57)

- There are 10 rubber ducks in a box. How many ducks are in _____ boxes?

    (2)   (6)   (13)   (20)

- There were _____ fireflies in the night sky. Ben caught _____ of them. How many fireflies were left in the sky?

    (30, 10)   (20, 6)   (50, 25)   (52, 30)   (99, 25)

1.NBT: Use place value understanding and properties of operations to add and subtract.

- There were _____ tyrannosaurs and _____ triceratops in the valley. How many dinosaurs were there all together?

    (11, 10)   (22, 20)   (32, 43)   (40, 45)   (67, 88)

- There were _____ brontosaurs laying eggs. Each brontosaur laid _____ eggs. How many eggs did they lay in all?

    (10, 2)   (8, 5)   (7, 10)   (14, 5)   (7, 15)

- T. rex has really sharp teeth. He has _____ teeth on the top and _____ teeth on the bottom. How many teeth does T. rex have?

    (10, 17)   (24, 25)   (32, 30)   (46, 48)   (52, 62)

- Stegosaurus has 2 rows of plates on his back. Each row has _____ plates. How many plates does stegosaurus have?

    (8)   (11)   (24)   (42)   (55)

- Diplodocus ate _____ leaves off a tree. He found another tree and ate _____ more leaves. How many leaves did diplodocus eat?

    (20, 30)   (21, 31)   (36, 26)   (4, 48)   (54, 38)

2.NBT: Understand place value.

- Annie has _____ marbles in a jar. Grace has _____ marbles in a jar. How many marbles do Annie and Grace have?

    (65, 42)   (58, 102)   (179, 81)   (347, 263)   (206, 474)   (397, 603)

- Write a number sentence for each pair of numbers comparing Annie's marbles to Grace's marbles (use <, >, or =).

- The fruit bats were hungry. In the rainforest, there are 10 bananas in a tree. If there are _____ trees, how many bananas are there?

    (4)   (7)   (11)   (41)   (62)   (100)

- There are 5 Honduran white bats roosting under each leaf in the rainforest. There are _____ leaves. How many Honduran white bats are there?

    (3)   (7)   (8)   (10)   (11)   (15)   (20)

- Mr. Butz has _____ pencils. He wants to give the same number of pencils to each of his students. There are 25 students in his class. How many pencils did each student get?

    (50)   (100)   (175)   (225)   (350)   (500)   (1000)

2.NBT: *Use place value understanding and properties of operations to add and subtract.*

- There are _____ penguins at the zoo. How many more penguins need to come to the zoo to have _____ penguins?

A	(6, 10)	(12, 20)	(15, 30)
B	(60, 100)	(32, 50)	(150, 300)
C	(160, 300)	(99, 125)	(130, 250)
D	(600, 1000)	(249, 500)	(473, 801)

- One day Mrs. Franke's class went to an Iowa State Cyclones game. First there were _____ fans in the stadium. Then some more fans came. Now there are _____ fans in the stadium. How many fans came into the stadium?

A	(38, 50)	(89, 101)	(5, 101)
B	(63, 150)	(97, 175)	(120, 300)
C	(195, 456)	(279, 686)	(980, 1501)

- Matt has _____. He wants to buy a book that costs _____. How much more does he need to earn to buy the book?

A	(15¢, 25¢)	(25¢, 50¢)	(39¢, 80¢)	(51¢, $1.00)
B	(95¢, $1.25)	(43¢, $1.70)	(69¢, $3.00)	($1.40, $5.00)

First and Second Grade Band Problems for the Associative Property

- Brown squirrel collected _____ acorns on Monday, _____ acorns on Tuesday and _____ acorns on Wednesday. How many acorns did brown squirrel collect for winter?

    (3, 7, 4)   (4, 8, 4)   (9, 4, 5)   (2, 4, 6)

- A group of animal friends set off to see the world. There were _____ roosters, _____ cats, _____ dogs, and _____ turtles. How many animals went on the trip together?

    (2, 3, 7, 8)   (6, 7, 6, 7)   (10, 4, 12, 4)   (13, 13, 18, 21)

- My flower garden has _____ pink flowers, _____ purple flowers, and _____ white flowers. How many flowers are in my garden?

    (5, 4, 5)   (3, 7, 10)   (18, 6, 2)   (15, 5, 9)

- The fall tree had many beautiful leaves. There were _____ yellow leaves, _____ red leaves, _____ brown leaves, and _____ orange leaves. How many leaves were on the tree?

    (9, 7, 1, 3)   (11, 13, 19, 17)   (23, 19, 27, 31)

- The kids in room 206 were making snowballs. Johnny made 6. Alex made 10. Jessica made 6. Ellie made 4. The kids in room 208 were making snowballs too! Bryan made 12. Lindsay made 8. Sarah made 12. Which class made more snowballs?

Second and Third Grade Band Problems for the Associative Property

- There were _____ Honduran white bats, _____ flying fox bats, and _____ dog-faced bats at the zoo. How many bats were at the zoo?

  (10, 15, 12)   (26, 13, 14)   (19, 41, 36)   (25, 38, 25)

- Our class collected cans for the food drive. On Monday we collected _____ cans. On Tuesday we collected _____ cans. On Wednesday we collected _____ cans, and on Thursday we collected _____ cans. How many cans did we collect for the food drive?

  (13, 25, 17, 50)   (37, 50, 53, 50)   (75, 50, 75, 25)   (102, 88, 45, 45)

- The second grade classes went on a nature walk. Each class collected leaves. Room 205 collected _____ leaves. Room 206 collected _____ leaves. Room 207 collected _____ leaves. How many leaves did the third grade classes collect?

  (65, 145, 80)   (203, 113, 137)   (119, 201, 120)   (76, 125, 114)

- Abby had _____ seashells in her collection. She found _____ seashells at the beach. Allie had _____ seashells in her collection. She found _____ more at the beach. How many seashells do Abby and Allie have in their seashell collection?

  (48, 12, 36, 24)   (115, 30, 30, 45)   (26, 74, 51, 49)   (105, 55, 40, 99)

Appendix D: Pre-Assessment Chart

When using the chart on the following page, write the problem you will be posing in the labeled space. (See chapter 7 for support in choosing a problem and creating appropriate number choices for your students.) As students solve the problem, record their strategies in the labeled columns. After a student has solved the problem, ask him or her key questions, such as the ones below, to check for understanding:

- How did you decide to solve this problem? (Notice the structure the student used.)

- What does (mathematical symbol mean to you? (Use to gain additional information on child's understanding of symbols such as +, –, and =.)

- What is the solution to this problem? (Important because a child may "solve" a problem, but the process that the child orally describes may not match what they actually did.)

- How did that sound as you were counting? (This gives the student the opportunity to verbally count out the strategy he or she just used. There are times when the written or model representation does not match how the student counted.)

Problem:

Key questions:

Direct Modeling	Counting	Invented/Standard Algorithm—Correct	Invented/Standard Algorithm—Incorrect
Student and Details	Student and Details	Student and Details	Student and Details

References

Ball, A.F. "Toward a Theory of Generative Change in Culturally and Linguistically Complex Classrooms." *American Educational Research Journal*, 46, no. 1 (2009): 45–72.

Carpenter, T.P., E. Fennema, M.L. Franke, L. Levi, and S.B. Empson. *Children's Mathematics: Cognitively Guided Instruction*. Portsmouth, N.H.: Heinemann, 1999.

Carpenter, T.P., M.L. Franke, and L. Levi. *Thinking Mathematically: Integrating Arithmetic & Algebra in Elementary School*. Portsmouth, N.H.: Heinemann, 2003.

Drake, C., T.J. Land, T.G. Bartell, J.M. Aguirre, M.Q. Foote, A. Roth McDuffie, and E.E. Turner. "Three Strategies for Opening Curriculum Spaces: Building on Children's Multiple Mathematical Knowledge Bases While Using Curriculum Materials." *Teaching Children Mathematics* (in press).

Drake, C., T.J. Land, N. Franke, J. Johnson, and M. Sweeney. *Teaching Mathematics for Understanding.* (Reston, Va.: NCTM, forthcoming).

Empson, S.B. and L. Levi. *Extending Children's Mathematics: Fractions and Decimals: Innovations in Cognitively Guided Instruction*. Portsmouth, N.H.: Heinemann, 2011.

Fosnot, C.T. and M. Dolk. *Young Mathematicians at Work: Constructing Multiplication and Division*. Portsmouth, N.H.: Heinemann, 2001(a).

Fosnot, C.T. and M. Dolk. *Young Mathematicians at Work: Constructing Number Sense, Addition, and Subtraction*. Portsmouth, N.H.: Heinemann, 2001(b).

Fosnot, C.T. and M. Dolk. *Young Mathematicians at Work: Constructing Fractions, Decimals, and Percents*. Portsmouth, N.H.: Heinemann, 2002.

Jacobs, V.R., M.L. Franke, T.P. Carpenter, L. Levi, and D. Battey. "Professional Development Focused on Children's Algebraic Reasoning in Elementary School." *Journal for Research in Mathematics Education* 38, no. 3 (2007): 258–88.

Jacobs, V. R, L. Lamb, and R. Philipp. "Professional Noticing of Children's Mathematical Thinking." *Journal for Research in Mathematics Education* 41, no. 2 (2010): 169–202.

Lampert, M., H. Beasley, H. Ghousseini, E. Kazemi, and M. Franke. "Using Designed Instructional Activities to Enable Novices to Manage Ambitious Mathematics Teaching." In *Instructional Explanations in the Disciplines*, edited by M.K. Stein and L. Kucan. New York: Springer, 2010.

Land, T.J. and C. Drake. "Enhancing and Enacting Curricular Learning Progressions in Elementary Mathematics." *Mathematical Thinking and Learning* 16, no. 2 (2014): 109–34.

Molina, M., E. Castro, and R. Ambrose. "Enriching Arithmetic Learning by Promoting Relational Thinking." *The International Journal of Learning* 12, no. 5 (2005): 265–70.

Moses, R.P. and C.E. Cobb. *Radical Equations: Civil Rights from Mississippi to the Algebra Project.* Boston: Beacon Press, 2001.

National Council of Teachers of Mathematics. *Principles and Standards for School Mathematics.* Reston, Va.: NCTM, 2000.

National Governors Association Center for Best Practices and Council of Chief State School Officers (NGA Center and CCSSO). *Common Core State Standards for Mathematics. Common Core State Standards* (*College- and Career-Readiness Standards and K–12 Standards in English Language Arts and Math*). Washington, D.C.: NGA Center and CCSSO, 2010. http://www.corestandards.org.

Parrish, S. *Number Talks, Grades K–5: Helping Children Build Mental Math and Computation Strategies.* Sausilito, Calif: Math Solutions Publications, 2010.

Routman, R. *Reading Essentials: The Specifics You Need to Teach Reading Well.* Portsmouth, N.H.: Heinemann, 2003.

Russell, S.J. "CCSSM: Keeping Teaching and Learning Strong." *Teaching Children Mathematics* 19, no. 1 (2012): 50–56.

Stigler, J.W., and J. Hiebert. *The Teaching Gap: Best Ideas from the World's Teachers for Improving Education in the Classroom.* New York: Simon & Schuster, 2009.

TERC. *Investigations in Number, Data, and Space.* Glenview, Ill.: Pearson/Scott Foresman, 2008.

—— . *Investigations in Number, Data, and Space: Grade 3, Unit 1.* Glenview, Ill.: Pearson/Scott Foresman, 2012.

The University of Chicago School Mathematics Project (UCSMP). *Everyday Mathematics.* Chicago: McGraw Hill, 2007.

About the Authors

Tonia Land is assistant professor of mathematics education at Drake University in Des Moines, Iowa. Her research interests include preparing preservice teachers in the productive use of standards-based curricular materials, with a focus on responding to children's mathematical thinking.

Corey Drake is associate professor of mathematics education and director of teacher preparation at Michigan State University. Her research and teaching focus on the preparation of elementary teachers to teach mathematics to diverse groups of children, with special attention to children's mathematical thinking.

Molly Sweeney has been a classroom teacher for more than fifteen years, teaching students from second through fifth grades. She has also worked as an instructional math leader.

Natalie Franke has been a teacher for eighteen years in grades 2–4. Her career also includes experience as an assistant principal.

Jennifer M. Johnson is a classroom teacher with more than twenty-five years of experience in the primary grades. She is a recipient of a Presidential Award for Excellence in Mathematics and Science Teaching.

Sweeney, Franke, and Johnson are also Iowa Department of Education Cognitively Guided Instruction (CGI) Trainers. Trainers lead CGI professional development for local school districts in the state.